PATRICK MOORE
ON
MARS

PATRICK MOORE
ON
MARS

CASSELL

Cassell
Wellington House, 125 Strand
London WC2R 0BB
www.cassell.co.uk

Distributed in the United States by
Sterling Publishing Co. Inc.
387 Park Avenue South
New York NY 10016-8810

British Library Cataloguing-in-Publication Data
A catalogue record for this book is available from the
British Library

ISBN 0-304-35069-9

Designed by Roger Chesneau

Printed and bound in Great Britain by
Creative Print and Design (Wales), Ebbw Vale

Contents

Preface

When I first wrote my book *Guide to Mars*, in 1955, our ideas about the Red Planet were very different from those of today. We believed the dark areas to be due to vegetation, and that even the 'canals' had a basis of reality, though the idea of intelligent Martians had been abandoned.

Today we have found that Mars is a world of volcanoes, valleys and craters; whether or not there is a trace of life there we still do not know, though we are sure that there is nothing like so advanced as a blade of grass. Space-craft have been there, and have sent back images direct from the surface; as I write these words one probe is circling the planet, and another is on its way there.

The last edition of my Guide came out over twenty years ago, so that this is a new book; only some of the historical sections have been retained. No doubt it too will be out of date in the near future, but meantime I hope that it will be of interest to those who read it.

My grateful thanks go to Dr. Peter Cattermole, for checking parts of the manuscript; to the authorities at the Jet Propulsion Laboratory and at NASA for permission to use the pictures sent back by the space-probes; to Barry Holmes, of Messrs. Cassell, for his constant help and encouragement; and to Roger Chesneau, for his great help in checking and proof-reading.

PATRICK MOORE
Selsey, September 1998

CHAPTER 1

The Red World

That Mars is inhabited by beings of some sort or other we may consider as certain as it is uncertain what those beings may be.

So wrote the American astronomer Percival Lowell in his book *Mars and its Canals*, published in 1906. He painted a vivid picture of a world which was desperately short of water, and which was inhabited by a brilliant civilization capable of building a planet-wide irrigation system to convey the precious water from the polar ice-caps through to the warmer regions closer to the equator. His views were controversial even at the time, but all things considered there seemed to be no valid reason why Mars should not support life of reasonably advanced type.

We know better now. There are no Martians, and the canal network has been relegated to the realm of myth. Yet we still cannot say definitely that there is no life there at all, and it is certainly true that Mars is much less unlike the Earth than any other planet in the Solar System. Moreover, it is within our range. Unmanned space-craft have landed there, and have crawled around the surface under the control of operators millions of miles away; plans for a full-scale expedition there are already being worked out. It is quite likely that the first man on Mars has already been born.

My own observations of Mars go back to the year 1934, when I was eleven years old and had just acquired my first telescope (a 3-inch refractor, which I still have and which I still use). I was fascinated by the tiny red disk, with its dark patches and its white polar cap. Like most other people, I believed the dark markings to be due to vegetation, and that Mars was a living world. Thirty years later we could show that the vegetation-tracts are as unreal as Lowell's canals, but they were replaced by huge craters, deep valleys, ancient riverbeds, flood-plains and magnificent volcanoes. Despite the loss of its canal-builders, Mars was as intriguing as ever.

I remember that during the early 1960s I gave a lecture about Mars to an audience at the University of London. I made a dozen positive statements, all of which were supported by the best available evidence – and all of which turned out to be wrong. Mars is not the sort of world which we believed it to be before the space missions flew. My aim now is to present a concise, up-to-date account of the planet, and though I may still be wrong in some details I honestly do not think that I will be as wide of the mark as I was less than forty years ago. First, let me give a few facts and figures, so that Mars can be put into its proper place in the Solar System.

Everyone knows that the Earth is a planet, moving round the Sun at a mean distance of 93,000,000 miles in a period of one year. Almost everyone knows that the Sun is an ordinary star, and that all the stars visible on any clear night are themselves suns, many of them much larger, hotter and more luminous than ours. They look so much less splendid only because they are so much further away. The nearest star beyond the Sun – a dim red dwarf known as Proxima Centauri – is over 24 million million miles from us. Faced with distances of this order, the mile and the kilometre become inconveniently short, and instead we use the light-year, which is the distance travelled by a ray of light in one year; rather less than 6 million million miles. Proxima, then, is just over 4 light-years away. Most stars are much more remote than this; for example the distance of the Pole Star is 680 light-years, so that we see it now as it used to be in the time of the Crusades. Deneb, the bright star in the constellation of the Swan, is so remote that we are seeing it as it used to be when Britain was occupied by the Romans. And when we consider other star-systems, we have to reckon in millions, hundreds of millions and even thousands of millions of light-years. Once we look beyond our own local area, our view of the universe is bound to be very out of date.

Things are different within the Solar System, which is ruled by the Sun, and whose main members are the nine planets: Mercury and Venus, closer to the Sun than we are, and Mars, Jupiter, Saturn, Uranus, Neptune and Pluto, which are further away. The Sun itself is a huge globe of incandescent gas, big enough to swallow up more than a million globes the volume of the Earth; its diameter is around 865,000 miles, and it is shining because of nuclear processes going on deep inside it. The planets have no light of their own, and shine only because they reflect the rays of the Sun; this is also true of the Moon, which is the Earth's satellite and which stays together with us

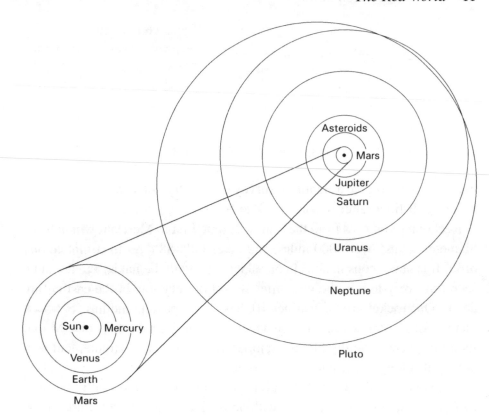

Orbits of the planets.

as we journey round the Sun. The Moon, a mere 239,000 miles from us, is much the nearest natural body in the sky. Light can travel from the Earth to the Moon in less than two seconds, but light from the Sun takes 8.6 minutes to reach us, and the outermost planets are about five 'light-hours' away.

So far as space-craft are concerned, we have to concede that we are limited to the Sun's family. True, several probes have by-passed the outer planets and are now leaving the Solar System for ever, but there is not the slightest chance of our being able to keep in touch with them for more than a few years. If we are ever to reach other planetary systems – which must surely exist – we will have to do so by some method which is so far beyond our present capability that we cannot even begin to speculate about it. We may be as far away from interstellar travel as King Canute was from television.

A question which I am always being asked is: 'What are the chances of finding intelligent life within range of us?' So far as the Solar System is concerned, I fear that the answer must be 'Nil', and to drive this home it

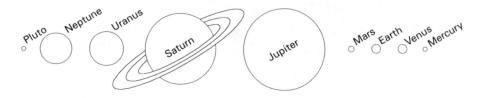

Comparative sizes of the planets.

may be a help to give a lightning survey of the other planets, after which it will be a distinct relief to come back to Mars.

Reckoning outward from the Sun, we come first to Mercury, which has a diameter of just over 3000 miles, and takes only 88 days to complete one orbit. It is never conspicuous in our skies, and with the naked eye it can be seen only very low in the west after sunset or very low in the east before dawn. One rocket probe, Mariner 10, has flown past it, and has sent back pictures of a crater-scarred surface which is not very unlike that of the Moon. Its atmosphere is negligible; the temperatures there are extreme, and as a potential colony it can safely be ruled out.

Venus is very different. It is about the same size as the Earth, and it is 67,000,000 miles from the Sun, with an orbital period of 225 Earth days. We might expect it to be reasonably welcoming, but before the Space Age we knew little about it, because it is permanently shrouded in a dense cloudy atmosphere which our telescopes cannot pierce. There were suggestions that the surface might be largely ocean-covered, in which case there seemed no obvious reason why life should not have started there – just as happened in the warm oceans of Earth thousands of millions of years ago.

Space missions proved that this is not so. The first probes found that the temperature at the surface is much too hot for any water to exist; radar images showed craters there, and in 1975 the Russians achieved a notable triumph by soft-landing two probes and obtaining the first pictures direct from the surface. Between 1990 and 1994 the American radar-carrying space-craft Magellan circled Venus, mapping the entire surface and showing plains, valleys, highlands, lava-flows, and volcanoes which are almost certainly active. Instead of being a friendly place, Venus turned out to be what can only be termed a psychedelic planet. The atmosphere is composed mainly of carbon dioxide, while the clouds are rich in sulphuric acid; the ground atmospheric pressure is about 90 times that of the Earth's air at sea-level, and the temperature is over 900 degrees Fahrenheit. Anyone incautious enough to go to Venus and step outside his space-craft would at once be fried, poisoned, squashed and corroded. It is not an inviting prospect, and I very much doubt whether any astronauts will try to land there in the foreseeable future.

Number three in the planetary sequence is the Earth, attended by the Moon. I do not propose to say a great deal about the Moon here, but I cannot gloss over it completely, because of inevitable comparisons with Mars; after all, both have cratered surfaces, together with high mountains and deep valleys. But the two worlds differ in many important ways, and comparisons cannot be taken too far.

The Moon is, on average, 239,000 miles from us, and has an orbital period of 27.3 days. Its diameter (2160 miles) is more than one-quarter that of the Earth. I suspect that the Earth–Moon systems should be regarded as a double planet rather than as a planet and a satellite, but the Moon has only 1/81 of the Earth's mass, and its gravitational pull is too weak for it to retain an appreciable atmosphere; there is no longer any doubt that it has been sterile throughout its long history.

The lunar surface is dominated by the walled formations known as craters, some of which are well over 100 miles across and have lofty central peaks. Most astronomers now believe that they were formed by meteoritic impacts, and that there was a period of intense bombardment, after which the lower-lying areas were flooded with lava welling up from below the crust; the 'seas' are very prominent indeed, but there has never been any water in them. Impact craters are only to be expected, and we also have good examples here on Earth, notably the famous Meteor Crater in Arizona and the smaller but almost equally perfect Wolf Creek Crater in Australia. Mars has plenty of impact formations, as do the icy satellites of the giant outer planets.

The Moon's rotation period of 27.3 days is exactly the same as the time which it takes to complete one orbit, so that the same face of the Moon is always turned toward us. Before 1959, when the Russians sent their unmanned probe Lunik 3 on a 'round trip', we had no direct knowledge of the averted hemisphere, but, predictably, it has been found to be just as cratered, just as sterile and just as mountainous as the side we have always known. The 'edges' of the Earth-turned hemisphere are very foreshortened, and difficult to map except from space vehicles. Also, some of the craters near the Moon's poles are so deep that their floors never receive any sunlight at all.

The United States Apollo programme began in the 1960s, and in July 1969 Neil Armstrong and Buzz Aldrin landed successfully in the waterless Sea of Tranquillity. Armstrong's 'one small step' was indeed a major turning-point in the history of mankind; the gap between the two worlds had at last been bridged. The Apollo programme ended in 1972, with No. 17 in the series, but since then there have been several unmanned missions; the most recent of these, Lunar Prospector, entered a closed orbit round the Moon in January 1998 and began a prolonged programme of mapping and analysis. There were claims that Prospector had confirmed a report from an earlier unmanned probe, Clementine, that ice existed inside some of the deep polar craters, whose floors are always bitterly cold. I admit that I am profoundly sceptical about this, but we must await the results of future research.

Originally it was thought that the Earth and the Moon used to be one body, and that the Moon was literally thrown off because of the rapid axial rotation. This idea was proposed by G. H. Darwin (son of the great naturalist) and was popular for many years before the mathematicians attacked it and more or less destroyed it. It is now widely believed that the two bodies were indeed one in the early days of the Solar System, and that a massive impact produced débris from which the Moon condensed. Certainly the Earth and the Moon are of about the same age – 4.6 thousand million years* – and no doubt this is also true of Mars. The fact that Earth, Mars and the Moon have evolved differently is because their masses are different. In particular, the Earth has retained a dense atmosphere, Mars a thin one, and the Moon virtually none at all.

Mars is the outermost of the inner or terrestrial group of planets, all of which are of modest size and have solid globes. Beyond we come to the minor planets or asteroids, all of which are below 600 miles in diameter and none of which can retain appreciable atmospheres; only one (Vesta) is ever visible with the naked eye. The idea that they are due to the breaking-up of a former planet has fallen into disfavour, and it now seems that no large planet could form in this region of the Solar System because of the powerful disruptive influence of Jupiter. Thousands of asteroids have now had their orbits worked out, and the total membership of the swarm may be as much

---

* I always avoid using the term 'billion', because the English and American billions are not the same.

as 50,000. They are relevant here because it is very likely that Phobos and Deimos, the two midget satellites of Mars, are ex-asteroids which were captured long ago. Positive proof is lacking, but the idea does seem very reasonable, and certainly Phobos and Deimos are quite different in nature from our massive Moon. It is also worth noting that one asteroid, No. 5261 Eureka, moves in the same orbit as Mars, though it keeps prudently well away from the planet and is in no danger of being engulfed. (It would not be at all prominent in the Martian sky, since it is less than two miles in diameter.)

Outside the main asteroid zone come the giant planets. Jupiter, moving round the Sun at an average distance of 483,000,000 miles, is almost 89,000 miles in diameter, and has a mass over 300 times greater than that of the Earth. Yet its density is relatively low – only 1.3 times that of water – and its structure is quite unlike those of the terrestrial planets. There is a silicate core, surrounded by layers of liquid hydrogen which are in turn overlaid by the dense, cloudy 'atmosphere' which we can see. It has a whole family of satellites, of which four are bright enough to be seen with any small telescope, and three of which are larger than our Moon.

Saturn, at a mean distance of 886,000,000 miles from the Sun, is also a gas-giant, and is distinguished by its superb system of rings, composed of icy particles moving round the planet in the nature of miniature moons. Saturn has one satellite, Titan, which is actually larger than the planet Mercury, and has a dense atmosphere composed largely of nitrogen, though the chances of life there seem to be slender in the extreme. Still further out come Uranus and Neptune, largely 'icy' and again with cloudy upper atmospheres. Uranus can just be seen with the naked eye if you know where to look for it, but Neptune is much fainter.

Various unmanned space-craft have been sent past the giant planets. Voyager 2, launched in 1977, went on what has been called a 'grand tour', encountering Jupiter (1979), Saturn (1981), Uranus (1986) and finally Neptune (1989), after which it began a never-ending journey out of the Solar System altogether. Superb images were obtained of all the giants and most of their satellites.

Pluto, discovered by Clyde Tombaugh in 1930, is not a giant; indeed it is smaller than the Moon, and also smaller than some of the planetary satellites. It has a curiously eccentric orbit, and travels in company with a second body, Charon, whose diameter is more than half that of Pluto.

Frankly, Pluto is an enigma. It does not seem to fit into the general pattern, and it may not be worthy of true planetary status. In recent years numbers of small, asteroidal bodies have been found in these remote regions, and it may well be that Pluto is merely the brightest member of an outer-asteroid swarm.

Comets are the most erratic members of the Solar System, and are not nearly so important as they sometimes look. Basically a comet is a 'dirty ice-ball', no more than a few miles or at most a few tens of miles across; when it draws in toward the Sun the ices begin to evaporate, and the comet develops a dusty head or coma, sometimes with a tail or tails. There are many small comets which move round the Sun in short periods, usually in highly eccentric orbits; really large comets come from the depths of space, and are apt to take us by surprise. Our last brilliant visitor was Comet Hale-Bopp of 1997, which shone magnificently in our skies for months. Unfortunately it will not be back for over 2000 years, and there is only one bright comet (Halley's) which has a reasonably short period; it last came back in 1986, and will be with us once more in 2061. However, we may have another really spectacular comet at any moment. Some, such as the Great Comet of 1882, have been known to cast shadows.

As a comet moves, it leaves a dusty trail behind it. If the Earth passes through one of these trails, the particles dash into the upper air, and set up so much heat by friction that they burn away in the streaks of luminosity which we call shooting-stars or meteors. Many meteor showers occur every year, and there are also 'sporadic' meteors, which may come from any direction at any time. Meteors burn out at altitudes of about 40 miles above ground level, where the air is very thin; there is no reason to doubt that meteors will also be seen burning away in the upper atmosphere of Mars.

I appreciate that this is a very sketchy account of the Solar System, but at least it may help to set the scene for our main subject. Meanwhile, it is amusing (and instructive) to look back at the Guzman Prize, which was announced in Paris on 17 December 1900. It was quite substantial – a hundred thousand francs – and was to be given to the first man who was able to establish contact with a being living on another world. But in the terms of the citation, Mars was specifically excluded, because it was thought that contacting the Martians would be much too easy!

The Guzman Prize remains unclaimed, and there are, alas, no Martians. Luckily, the interest of the Red Planet remains as strong as ever, and there can be no world more intriguing for Man to explore.

CHAPTER 2

Mars in the Solar System

It was once thought that the planets must move in circular orbits. This was because the circle was regarded as the perfect form – and surely nothing short of absolute perfection could be allowed in the heavens? This may not seem a very scientific way of looking at things, but in bygone ages it carried a great deal of weight.

I do not want to delve too deeply into history, but since Mars comes very much into the story it is worth digressing for a few moments. Ptolemy of Alexandria, the last great astronomer of Classical times, followed a scheme according to which the Earth lay at rest in the centre of the universe, with everything else moving around it; the Moon, Sun, planets and, at the very edge, the sphere of the fixed stars. Ptolemy, who was an excellent mathematician, knew quite well that the planets do not move across the sky at a steady rate, and so he adopted a rather cumbersome scheme: a planet moved in a small circle or epicycle, the centre of which (the deferent) itself moved round the Earth in a perfect circle. This sounds artificial, but it did fit the facts as Ptolemy knew them. The scheme is always known as the Ptolemaic System, though in fact Ptolemy did not invent it; he merely brought it to its highest degree of accuracy.

Ptolemy died in or around AD 180, and for centuries after his death very few people dared to question his model of the universe. Religion was very much to the fore. The Church was strongly against any heretical proposal to dethrone the Earth from its proud central position, and to question the authority of the Church was decidedly unwise. The real battle started in the fifteenth century, with the publication of a book by a Polish canon who is always remembered as Copernicus, though his proper name was Mikołaj Kopernik. Copernicus objected to the Ptolemaic System because it was over-complicated, and his remedy was to remove the Earth from its central position and put the Sun there instead. This would mean relegating the Earth to the status of an ordinary planet, no more important, basically, than Mars.

Church reaction was predictably violent, and Copernicus only escaped severe persecution because he was prudent enough to postpone publication of the book until he was dying. Subsequently the controversy became over-heated by any standards, and one rebel, Giordano Bruno, was burned at the stake in Rome in 1600, partly because he refused to believe that the Sun goes round the Earth.

Meantime, a curious character named Tycho Brahe had come into the story. He was a Danish nobleman, and he had a remarkable and picturesque life. This is no place to describe it in detail, or to relate the numerous stories about him – though I cannot resist recalling that during his student days he had part of his nose sliced off in a duel, and made himself a new one out of gold, silver and wax! He was a superb observer, and between 1576 and 1596 he worked away on his island observatory in the Baltic, preparing a new catalogue of star positions which was far better than anything previously drawn up. I am sure that I need not stress that so far as naked-eye observers are concerned, the stars keep to the same relative positions for long periods (many lifetimes), because they are so far away that their individual or 'proper' motions are very slight. Go back to the days of the Crusades, the Roman Occupation or even the Trojan War, and you would find that the constellations would look virtually the same as they do today. Orion, the Great Bear and the rest would be sensibly unaltered. Neither will they have changed perceptibly by, say, AD 2200. It is only our near neighbours – the members of the Solar System – which seem to wander about.

Tycho, unfortunately, had no telescopes. The first definite astronomical use of the telescope dates only from 1609, which was eight years after Tycho died. So the great Dane had to depend upon measuring instruments used with the naked eye alone, and under the circumstances he was amazingly successful. When he left Denmark in a huff, after quarrelling with practically everybody, he took with him a mass of observations. In addition to his star catalogue, he had made very exact studies of the movements of the planets, particularly Mars.

Tycho settled in Bohemia as Imperial Mathematician to the Holy Roman Emperor, Rudolph II, who was a gloomy, astrology-ridden individual, and whose régime was one of unmitigated disaster. It was during his spell in Bohemia that Tycho engaged as his assistant a young German, Johannes

Kepler, and when Tycho died suddenly in 1601 it was Kepler who came into possession of all the priceless observations.

This was a great opportunity. Tycho had never accepted the Copernican theory, and had developed a sort of hybrid variation of his own which met with little support (it has to be added that Tycho was also a firm believer in astrology). But Kepler believed in a central Sun, and with Tycho's work he was confident that he could prove it. It took him a long time, and it was Mars which provided the essential clue. If Mars moved round the Sun, then so must the Earth – but no circular orbit seemed to fit the observations.

Kepler persisted. Blind alley followed blind alley, but at last he stumbled upon the truth. Mars does indeed go round the Sun, but its path is not circular; the planet moves in an ellipse. Once this had been established, everything clicked neatly into place. The main battle was over, though the last shots remained to be fired; Galileo, the first great telescopic observer, was tried and condemned by the Roman Inquisition in 1633 because of his refusal to accept the Ptolemaic System, and it was not until 1687, with the publication of Isaac Newton's *Principia*, that the matter was finally closed. But it was Kepler who had provided the vital evidence.

I say that Kepler 'stumbled upon the truth' because he had had the solution in his grasp for some time without realizing it; to break free from the old concept of circular orbits required a great deal of moral courage. Incidentally, it was pure luck that Tycho's observations had been concentrated upon Mars, because the orbit of Mars is much more eccentric than those of, say, Venus or Jupiter. Venus really does have a path which is almost circular, and if Mars had behaved in the same way the solution would have been much harder to find. On the other hand, we must not belittle Kepler. He had the utmost faith in Tycho's observations, and if he had been less confident about them he would inevitably have failed.

In fact Mars has an orbital eccentricity of 0.093, as against 0.017 for the Earth and only 0.007 for Venus. This means that its distance from the Sun ranges between 128,410,000 miles at perihelion (the closest point to the Sun) and 154,860,000 miles at aphelion (furthest recession) – a difference of over 25,000,000 miles, as against less than 4,000,000 miles for the Earth. The mean distance between Mars and the Sun is therefore 141,700,000 miles. Note too that it is only the second closest planet to the Earth. It may come within 35,000,000 miles of us, but this is more than ten million miles further than Venus at its nearest.

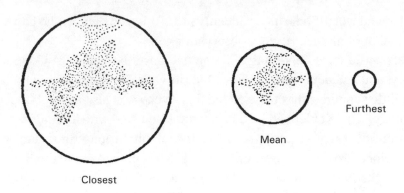

Closest

Mean

Furthest

Varying apparent size of Mars.

When well placed, Mars is the brightest object in the entire sky apart from the Sun, the Moon and Venus. It stands out not only because of its brilliance, but also because of the strong red colour which led the ancients to name it in honour of the God of War. Now and then it has even caused general alarm, as it did in 1719, when it was at its very best and many people mistook it for a red comet which was about to collide with the Earth. At its faintest it is not much brighter than the Pole Star, and at such times it is not easy for the beginner to identify it. As seen from Earth, the apparent diameter ranges between 25.7 seconds of arc and only 3.5 seconds of arc.

Mars takes 687 Earth days to go round the Sun. Obviously it is at its best when opposite to the Sun in the sky, but these 'oppositions' do not occur every year; the mean interval between them – known as the synodic period

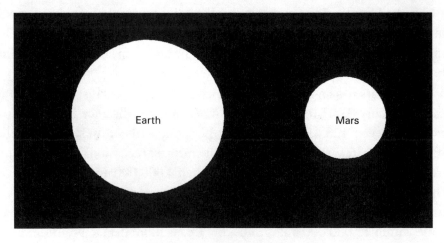

Earth

Mars

Earth and Mars compared.

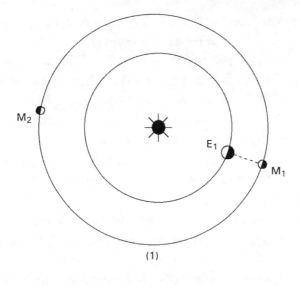

(1)

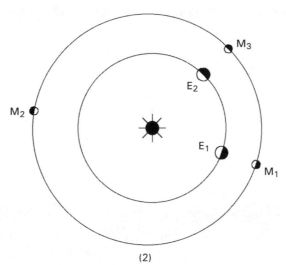

(2)

Oppositions of Mars.

– is 780 days, though this is not absolutely constant. To show what happens, let us refer to the diagram above, in which the orbits of the two planets are drawn to the same scale.

It will be convenient to begin on 22 September 1988, because on that date Mars was an opposition (M1 in the upper diagram). It was also practically at perihelion, so that it was almost as close to the Earth as it can ever be; it was glaringly conspicuous, and was visible throughout the hours of darkness. One year later the Earth had completed a full journey round the Sun, and had arrived back at its starting-point (E1), but Mars, moving more slowly in a larger orbit, had not had time to do so. Its mean orbital velocity is only 15 miles per second, as against 18½ miles per second for the Earth. By September 1989 Mars had only reached position M2, and was almost behind the Sun. The Earth had to catch it up, and it achieved this only on 20 November 1990, when there was another opposition, with the Earth at E2 and Mars at M3 (lower diagram). Throughout the first part of 1990 Mars was badly placed for observation, and small telescopes would show almost nothing on its disk.

This is why oppositions take place only in alternate years. Thus there will be oppositions in April 1999, June 2001 and August 2003, but none in 2000 or 2002.

Moreover, oppositions are not equally favourable. That of 2003 will take place with Mars near perihelion, and the apparent diameter will exceed 25

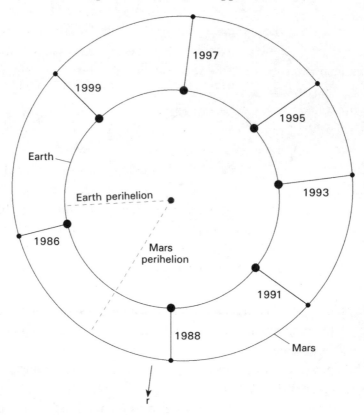

Oppositions of Mars 1986–99.

seconds of arc, but in 1995 the minimum distance was over 60,000,000 miles, with a greatest apparent diameter of less than 14 seconds of arc. A list of recent and future oppositions is given in Appendix II.

Because Mars shines by reflected sunlight, it shows a distinct phase. At times of opposition the whole of the sunlit face is turned toward us, and Mars is full, but when well away from opposition the disk no longer appears circular, and at times only 86 per cent of the daylight hemisphere is turned in our direction. Sketches made at the telescope have to take this into account, because if the disk is drawn as circular and the real phase is below around 95 per cent there are bound to be errors. Only when the planet is reasonably close to opposition can the careful observer afford to draw the disk as perfectly circular. The disks given on pages 24 and 25 may be of help.

This may be the moment to introduce the Martian seasons, which are all-important in any discussion of the surface conditions. As a preliminary I must say something about or own seasons, which are due in the main not to our changing distance from the Sun, but to the tilt of the Earth's axis, which is 23.4 degrees from the perpendicular to the orbit. In the next diagram (below), the Earth is shown in two positions. During northern summer, the northern hemisphere is tilted toward the Sun; for a while the north pole is in continual sunlight, while the south pole has no daytime at all. Six months later, the situation is reversed. The Earth is at perihelion during northern winter – 91½ million miles from the Sun, as against 94½ million miles in late June – but the difference is not enough to have any marked effect, and in any case the greater area of ocean in the southern part of the Earth has a stabilizing influence.

Not so with Mars, which has no oceans at all, and where the distance range between perihelion and aphelion is so much greater. At the present epoch, the axial tilt is virtually the same as ours (23 degrees 59 minutes), and, as with us, the southern summer occurs at perihelion. This means that on Mars the southern summers are shorter but hotter than those in the north-

The seasons on Earth.

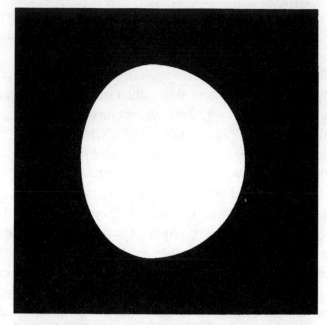

85%

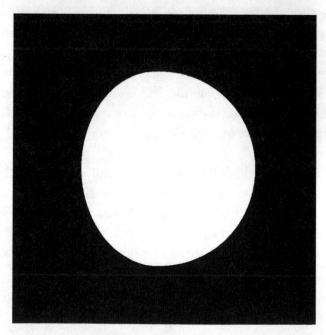

90%

When Mars is away from opposition it shows a definite phase, and this should be allowed for when making drawings.

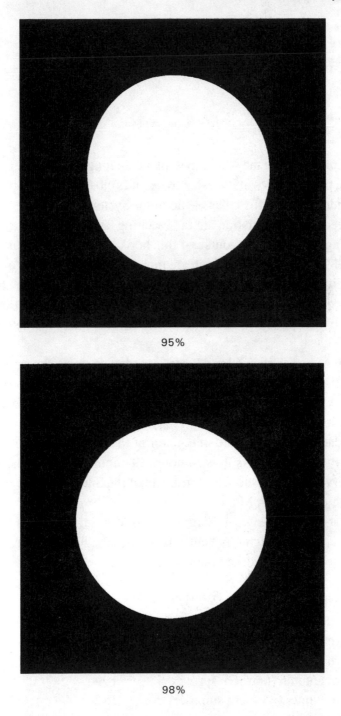

95%

98%

The outlines given here can be photocopied and used in sketching.

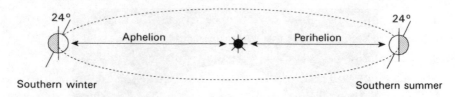

24°   24°

Aphelion   ☀   Perihelion

Southern winter   Southern summer

The seasons on Mars.

ern hemisphere, while the southern winters are longer and colder. As Kepler showed, a planet moves at its fastest when it is closest to the Sun; this is one of the fundamental traffic rules of the Solar System.

The effects of this situation are far-reaching, and I will refer to them over and over again during the course of this book. In short: the southern hemisphere of Mars has climates which are more extreme than those of the north, and the differences are quite marked. It is also worth noting that during perihelic oppositions it is the south pole which is tilted in our direction, so that before the Space Age the southern hemisphere was the better-mapped of the two.

Though Mars has such a long 'year', its 'day' is much the same as ours – to be precise, a little more than half an hour longer: 24 hours 37 minutes 22.6 seconds, now usually called a 'sol'. Of course, a sol indicates a complete rotation, with approximately 12¼ hours of daylight followed by 12¼ hours of darkness at the Martian equator at the time of the equinoxes. The solar day (or, rather, solar sol) is 24 hours 39 minutes 35 seconds. This is the mean interval between successive transits of the Sun across the meridian as seen by an observer on Mars.

The near-equality between Earth day and Martian sol means that we can draw up a calendar which is not entirely unfamiliar. For instance, the lengths of the seasons can be worked out easily enough:

Martian season	Length	
	Earth days	*Martian sols*
Southern spring (northern autumn)	146	142
Southern summer (northern winter)	160	156
Southern autumn (northern spring)	199	194
Southern winter (northern summer)	182	177
	687	669

Timekeeping is not going to be a real problem for future colonists (as it would be on Venus, where the rotation period or 'day' is actually longer than the revolution period or 'year', leading to a very peculiar state of affairs). A Martian clock will no doubt be divided into 24 hours, though each will be slightly longer than an Earth hour. So far as a calendar is concerned – well, there have already been plenty of suggestions. My own idea is to divide a Martian year up into 18 months, each of 37 sols. This makes 666 sols. We need 669, so an extra sol can be tacked on to Months 6, 12 and 18, giving them 38 sols instead of 37. This should work reasonably well, though eventually there will have to be some sort of adjustment to allow for the fact that the orbital period of Mars is not exactly 687 days, but 686 days 23 hours 52 minutes 31 seconds according to Earth reckoning. The problem of naming our eighteen Martian months is something which need hardly be considered yet, though no doubt it will lead to the usual tedious international squabbles when it does have to be tackled.

Incidentally, the north pole star of Mars is not the same as ours. Northward, the Earth's axis points to a position in the sky very close to Alpha Ursæ Minoris – Polaris, which seems to remain almost stationary with everything else moving round it. (I am, of course, talking from the viewpoint of a northern-hemisphere observer. Australians and South Africans can never see Polaris, and there is no bright south pole star; the nearest naked-eye object is the obscure Sigma Octantis, which is none too easy to see without using binoculars or a telescope.) Mars is well served. The north pole star there is Deneb or Alpha Cygni, which is of the first magnitude and will be a splendid navigational aid for future explorers. This may be just as well. We now know that Mars does have a magnetic field, but it is extremely weak,

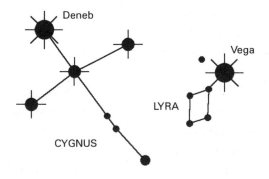

Deneb – the north pole star of Mars.

and we may expect old-fashioned magnetic compasses to be very unreliable!

However, Deneb will not retain its title indefinitely, because the axial tilt of Mars is not constant. The same is true of Earth, because of the phenomenon known as the precession of the equinoxes. Our world is not a perfect sphere; it is somewhat flattened at the poles, and the equatorial diameter is 26 miles greater than the polar diameter (7926 miles, as against 7900). The Sun and Moon pull upon the equatorial bulge, and so the axial tilt alters; the north celestial pole describes a circle in the sky over a period of 26,000 years, so that in 12,000 years' time the pole star will be the brilliant blue Vega rather than Polaris. Mars is flattened to a greater extent than the Earth; the discrepancy amounts to around 40 miles, and so the precessional effects are more marked even though Mars has no satellite massive enough to produce any measurable effects. There are various factors to be taken into account, but the sum total is that the axial tilt ranges between 14.9 degrees and 35.2 degrees over a cycle of 51,000 Earth years. In 25,000 years from now it will be the northern hemisphere of Mars which will have its summer at perihelion – and at perihelion Mars receives 45 per cent more radiation than it does when furthest from the Sun.

The weakness of the magnetic field brings us on to the question of what Mars is like inside its globe. The mean density is less than that of the Earth; what we call the specific gravity is only 3.94, so that Mars 'weighs' 3.9 times as much as an equal volume of water would do. For Earth, the figure is 5.5, due largely to the fact that there is a substantial core made up of heavy substances such as iron. The Moon, with a specific gravity of 3.4, has a core probably about 360 miles in diameter, over which lies a region of 'partial melting' which is in turn overlaid by the mantle and finally by the relatively thin crust, which goes down to only a few miles or at most a few tens of miles below the lunar surface.

Mars is more lunar than terrestrial in structure. It has only one-tenth of the Earth's mass, and may have built up rather more quickly from the material in the solar nebula. The outer layers solidified to form rocks of various types; the lighter rocks floated to the top to make a crust, while the heavier ones, relatively rich in iron, heated up and sank to form a core. One estimate gives the diameter of the present core as about 1850 miles, with a 2200-mile mantle and a crust which can hardly be more than 120 miles thick.

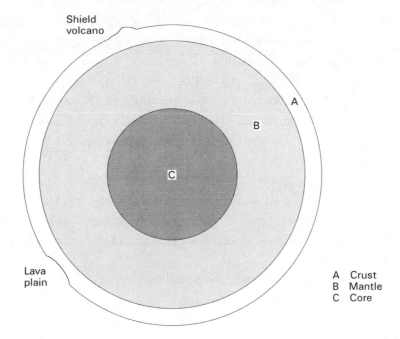

Internal structure of Mars

Come next to the important question of surface gravity. Again we may expect an intermediate value between those of the Earth and the Moon. The surface gravity of a planet (or a satellite) depends upon a combination of its mass and its diameter; the greater the mass and the smaller the diameter, the greater the surface pull. The reason why diameter is important is because a body behaves as though all its mass were concentrated at a point at its centre – and the further away you are from the centre, the weaker the gravitational tug. The giant planet Uranus is a case in point. Its diameter is 31,770 miles as measured through its equator, and its mass is well over 14 times that of the Earth. Anyone standing on Uranus would therefore be much further away from the centre of the planet than anyone standing on the surface of the Earth, so that despite the mass-difference the surface gravities on the two worlds would be much the same; if you weigh 11 stones on Earth, you would weigh just over 12 stones on Uranus. (Of course I am taking an impossible case, because Uranus has a gaseous surface and would be rather difficult to stand upon, but I think that the principle is clear.)

With the Moon, the surface gravity is about one-sixth of that of the Earth, so that an astronaut there will have only one-sixth of his Earth weight, and when he walks around everything seems to happen in slow motion. Most people will have seen this on television during the Apollo missions. Fortunately, it has been found that the sensation is neither uncomfortable nor dangerous, and the situation will be quite satisfactory on Mars, where the surface gravity is about one-third of that of Earth. (Note, by the way, that the surface gravity on Mercury is almost exactly the same as that on Mars. Mercury is much the smaller of the two planets, but it is also much denser, with a heavy, iron-rich core.)

Escape velocity also depends upon mass, but here the diameter is not so important when we are dealing with bodies of planetary size. By now 'escape velocity' has become part of our everyday language, but I think I must give a brief account of it, because it is so vital in any discussion about the conditions on Mars.

Throw an object upward, and it will rise to a certain height, stop, and fall back. Throw it harder, and it will travel further upward before returning. If it were possible to stand on the surface of the Earth and throw, say, a cricket ball upward at a speed of 7 miles per second, or 25,000 m.p.h., it would not fall back at all; the Earth's gravity would be unable to retain it, and the cricket ball would escape into space. Therefore, 7 miles per second is the Earth's escape velocity. Actually, any body moving through the dense lower atmosphere at 7 miles per second would burn away by frictional heating against the air-particles, which is one reason why the space-guns so beloved of science fiction writers will not work, but again the general principle is clear enough.

The Moon has an escape velocity of 1½ miles per second, and for Mars the value is 3.1 miles per second. (For Uranus, it is almost 14 miles per second.) And this has a marked effect upon the density of any remaining atmosphere.

Air is made up of vast numbers of tiny particles, all moving around at high speeds. Obviously, a particle moving outward from the Earth with a velocity of 7 miles per second will escape. Fortunately, our air is made up of particles which cannot attain this speed, and there is no danger of our being left gasping like a goldfish which has been removed from its tank. But the Moon, with its weak gravitational pull, has been unable to hold on to any appreciable atmosphere, which is why it is so unfriendly. Mars, predictably, is an intermediate case. It would be expected to have a thin atmosphere, and this is correct, though it is true that the atmospheric density is much less

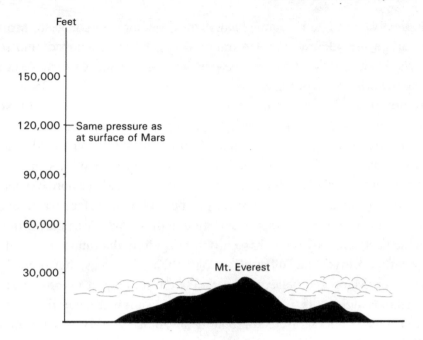

Atmospheric pressure on Mars

than we believed before the first rocket probes sent back information from close range.

Mars and the Earth must be of about the same age – 4.6 thousand million years – and no doubt they began to evolve along the same lines. Gases and vapours sent out by volcanic activity from below their newly-formed crusts produced atmospheres, and there was enough water vapour to fill the ocean beds (though more water may well have been added by impacting comets). In these early atmospheres hydrogen was abundant, which is no surprise; after all, hydrogen is the most plentiful substance in the entire universe, and the numbers of hydrogen atoms exceed those of all the other elements combined. But hydrogen is a light gas, and its quick-moving particles soon escaped. Heavier gases could not do so, because their particles moved more slowly.

Here, however, the stories of the Earth and Mars diverge. On Earth, the atmosphere became stable. Life began, plants spread on to the lands, and by the process of photosynthesis took in carbon dioxide and replaced it with free oxygen. (This was not a rapid process; even at the time of the dinosaurs, the atmosphere was so unlike that of today that we could not breathe it. Borrowing a time machine and travelling back into the Jurassic Period is emphatically not to be recommended.) On Mars, most of the gases apart from carbon dioxide escaped. No tracts of vegetation developed, and so there was no way to produce free oxygen. The oceans vanished, and Mars became the arid world of today.

The present atmosphere is made up chiefly of carbon dioxide, and is so thin that the ground pressure is everywhere below 10 millibars, as against almost 1000 millibars at sea-level on the Earth. This means that the Martian atmosphere is no denser than our air at a height of 15 miles above the ground, and we would be unable to breathe it even if it were made of pure oxygen. From this, we can also show that there can be no liquid surface water, and there are no seas, lakes or even ponds – though things were certainly different in the past, and may possibly be different again in the future.

The early observers had other ideas. With their telescopes, they drew the markings on Mars, and believed that they were looking at oceans, continents, islands and straits. It was only during the last few decades that a true picture began to emerge. But I am running ahead of my story, so let us now turn back the pages to those far-off times even before the telescopic exploration of Mars began.

CHAPTER 3

Eye and Telescope

Not even the most myopic observer can overlook Mars when it is best placed. Obviously, then, it must have been known since very early times, and it was one of the first bodies to be identified as a 'wanderer' rather than a star.

Astrology, the superstition of the stars, was once taken very seriously indeed, and Mars was generally held to have an evil influence. (This is no place to discuss astrology, which is, after all, strictly for the credulous; I have often commented that it proves only one scientific fact: 'There's one born every minute.') But it is understandable that our remote ancestors should have regarded Mars as baleful, because it is so red. Red indicates blood, and blood means war. What could be more natural than to name the planet after the War-God? The Greek war-god was Ares, and the study of Mars is still known officially as 'areography', though by now the term seems pedantic, and I doubt whether it will survive for much longer. 'Martian geography' is so much more explicit, though grammatically deplorable.

The Greeks were well aware of the unstarlike nature of Mars, and they studied its movements carefully, but they were not the first to do so. Apparently the Egyptians knew the planet as Harmarkhis or Har decher (the Red One), while the Chaldæans called it after Nergal, the Babylonian god of war. The first precise observation of its position seems to date from 272 BC, just over half a century after the death of Alexander the Great; on 17 January of that year it is recorded that Mars passed very close to the star Graffias or Beta Scorpii. Later on, the Arabs knew Mars as Mirikh, indicating a torch, while in India it was Angaraka, from *angara*, a burning coal.

In pre-telescopic times nothing could be learned about Mars itself. All that could really be done was to follow its movements against the starry background, and by the time that telescopes arrived on the scene the way in which Mars behaved was very well known – thanks largely, though by no

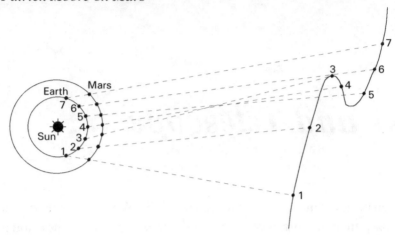

Retrograding of Mars. It is obvious that as the Earth 'overtakes' Mars, the planet will
seem to reverse its direction for an appreciable period before resuming its normal
direction of motion.

means entirely, to Tycho Brahe. Observers who are sufficiently interested
can easily plot the changing rate and direction of Mars as it shifts across the
night sky. Near opposition it will seem to move backwards, or east to west,
against the stars for some time, because the Earth is 'catching Mars up' and
passing it. The diagram should make this clear. The backwards motion is
known as 'retrograding', and was one of the phenomena which showed
from the outset that Mars could not move round the Earth in a circular path
at a steady rate.

The telescope was developed in the early years of the seventeenth century. There can be no doubt that a Dutch spectacle-maker named Lippershey
produced a working telescope in 1608, and he has been widely assumed to
have been the first, though some scientific historians have their doubts, and
some of them suspect that the first telescopes were made much earlier; since
spectacles had been in use for so many years, I have always wondered why
the discovery of the principle of the telescope was so long delayed.* Tho-
mas Harriott, one-time tutor to Sir Walter Raleigh, drew a telescopic map of
the Moon in 1609, and he was only one of several early observers, but pride
of place must undoubtedly go to the great Italian, Galileo Galilei, who made
a telescope for himself and began astronomical observing during the winter

---

* The British historian of astronomy C. A. Ronan maintained that a curious kind of telescope
was built in England by one Leonard Digges, at some time between 1550 and 1560. He may well
have been right, though we have no positive proof.

of 1609–10. At once he made a series of spectacular discoveries. Jupiter was found to be attended by four satellites; Venus showed phases like those of the Moon, and the Moon itself was a world of mountains and craters; there was something very odd about the shape of Saturn, and the Milky Way was found to be made up of myriads of stars. Galileo was quick to publish his work, and sparked off a bitter controversy, because everything that he saw strengthened his belief that the Sun, not the Earth, is the centre of the Solar System. The phases of Venus were of special importance here, because on the old Ptolemaic theory Venus would always show up as a crescent – never as a half or full disk.

There were two planets which puzzled Galileo. One, of course, was Saturn, which he first believed to be triple, though he later found that the two side members had disappeared. In fact this appearance was due to the ring system, which could be seen in Galileo's imperfect, low-powered telescope, but not clearly enough for him to make out what they were. (Their temporary disappearance was because in 1612 the rings were tilted edgewise-on to the Earth, as happens regularly; the last occasion was in 1995, the next will be in 2009.) Mars was certainly single, but it did not seem to be quite circular, and on 30 December 1610 Galileo wrote to his friend P. Castelli: 'I ought not to claim that I can see the phases of Mars; however, unless I am deceiving myself, I believe I have already seen that he is not perfectly round.' He was correct, and the fact that he could see the phases at all shows how good an observer he was.

What he could not do was to see any surface features on Mars, and this brings me to a point which deserves to be stressed. Mars is not an easy object to study; it is decidedly difficult, and small telescopes will show little on it even at the best of times. Unless a fairly powerful telescope is available, Mars will obstinately defy all attempts to tackle it.

There is a very good reason for this. Mars is a small world, and it is by no means on our cosmical doorstep, since it can never approach us much within 35,000,000 miles. This is around 150 times as far away as the Moon; even when at its nearest it has an apparent diameter of only 25.7 seconds of arc, and it would fit comfortably into one of the Moon's craters, such as Plato – even though Plato is a mere 60 miles across. No wonder that high magnifications are needed to show Mars well.

However, the early telescope-users were not to be deterred, and they did their best. The very first telescopic sketch of Mars seems to have been made

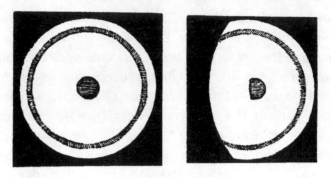

Drawings of Mars by F. Fontana: (left) 1636, (right) 1638.

in 1636 by an Italian amateur, Francisco Fontana, who lived in Naples; he had made his own telescope, which was rather better than Galileo's but was, naturally, very poor judged by modern standards. His drawing, shown here (above left), is hardly of great scientific value, but Fontana's own words are worth quoting: 'The form of Mars was observed to be perfectly spherical. In its centre was a dark cone in the form of a very black pill. The disk was of many colours, but appeared to be flaming in the concave part. Except for the Sun, Mars is much the hottest of all the stars.' A second drawing (above right), made two years later – on 24 August 1638 – showed the same 'pill', with a grossly exaggerated phase.

Fontana cannot be blamed for misinterpreting what he saw. The 'pill', of course, was purely an optical effect due to the poor quality of his telescope (he recorded a similar spot on Venus), and the 'many colours' were equally spurious, but at least he had made an effort to be constructive.

Passing over one or two earnest but frankly unsuccessful efforts to record detail on Mars during the years following Fontana's sketches, we come to Christiaan Huygens, of Holland. At that time the best available telescopes were small-aperture refractors of immensely long focal length. They were incredibly awkward to use, but Huygens was probably the best observer of the early seventeenth century, and he made several outstanding discoveries. For instance, he found out the true nature of Saturn's ring system, and was the first to see Titan, the largest and brightest of the Saturnian family of satellites. (He was also the inventor of the pendulum clock, and it is for this that he is best remembered.) He made a drawing of Mars in 1656, but it was on 28 November 1659 that he recorded the first genuine feature: the triangular dark patch which has been named, at various times, the Atlantic Canal,

Drawing of Mars by Huygens, 28 November 1659.
The Syrtis Major is unmistakable.

the Kaiser Sea, the Hourglass Sea, the Syrtis Major, and now, since the
space-probe surveys, the Syrtis Major Planitia.

The drawing was made at seven o'clock in the evening. The Syrtis is
unmistakable; true, its size is exaggerated, but of its identity there can be no
doubt at all. It is still there, and it is the most obvious dark patch on the
whole of the disk. Huygens' sketch, shown here (above), gives extra proof –
if proof were needed – that the Martian markings are permanent rather than
cloudlike, as with Venus. Moreover we can check back and see whether the
Syrtis Major really was in the position in which Huygens drew it on the
disk. The agreement is excellent, so that clearly the rotation period of the
planet has not changed.

Huygens looked at the Syrtis Major over a period of hours (at least, we
must so assume), and he saw that it was being carried slowly across the
disk. The direction would be from right to left as shown in the sketch, because the orientation is with south at the top.* The rate at which the mark-

---

* Most astronomical telescopes give an upside-down or inverted image, and before the Space
Age nearly all lunar and planetary images were shown with south at the top and north at the
bottom. The modern trend is to have north at the top. (This was the subject of a heated discussion
thirty years ago at a meeting of the International Astronomical Union, the controlling body of
world astronomy, and in a final vote the 'south-toppers' such as myself were heavily defeated!)
In the present book I am in something of a difficulty; it is sensible to show the drawings with
south uppermost, but the space pictures must be reversed. I hope that all this does not sound too
confusing.

ing shifts is a clue to the length of the rotation period. Huygens made no mistake, and in his diary for 1 December 1659 he recorded that 'the rotation of Mars, like that of the Earth, seems to be in a period of twenty-four hours'. This is a mere half-hour or so wrong, and with his limited equipment Huygens could scarcely have done any better. He must have had eyes like a lynx.

During his career Huygens made various other drawings of Mars, again showing the Syrtis Major, but he was not alone. Another great pioneer was Giovanni Cassini, Italian by birth but who spent much of his life in France, and became the first Director of the Paris Observatory. In 1666 (actually before he went to Paris) Cassini made several observations of Mars, and although his drawings were not so good as those of Huygens he found that the markings came back to the same positions on the disk forty minutes later each night – so that after 36 to 37 days they came back to the same positions at the same times. It followed that the rotation period of Mars must be slightly longer than that of the Earth, and Cassini arrived at a value of 24 hours 40 minutes, which is practically correct. There was some argument about this, and a 13-hour period was initially preferred by the telescope-maker Campani and his associates in Rome, but before long it became clear that Cassini was right.

It was also Cassini who was the first to record the white polar caps, in 1666. They looked so like the icy caps of Earth that their nature was not seriously questioned at the time, particularly as they were seen to wax and wane according to the Martian seasons. In 1704 and again in 1719 they were carefully studied by Giacomo Maraldi, Cassini's nephew, who gave a rotation period of 24 hours 39 minutes. Since the real period is just over 24 hours 37 minutes, Maraldi's value was very near the truth.

We come now to Sir William Herschel, one of the giants of astronomical history. He was Hanoverian by birth, but spent most of his life in England. In 1781, when he was carrying out a 'review of the heavens' with a home-made reflecting telescope, he discovered the planet Uranus; he catalogued thousands of star-clusters and nebulæ; he found that many of the double stars in the sky are physically-associated or binary pairs rather than mere line-of-sight effects, and he was the first man to draw up a reasonably good picture of the shape of the star system or Galaxy. He became the official astronomer to King George III of England and Hanover, and he received every honour that the scientific world could bestow, including a knighthood

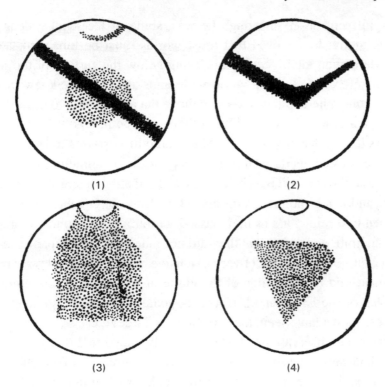

Drawings of Mars by Maraldi, 1719: (1) 13 July; (2) 19 August, 25 September and 28 October; (3) August–October; (4) 5 August and 16 October.

and the Presidency of the newly-formed Royal Astronomical Society (then the Astronomical Society of London). He became much the best telescope-maker of his time, and his largest telescope, a reflector with a 49-inch mirror, was much the most powerful ever made. You can see the mirror if you go to Flamsteed House, the Old Royal Observatory in Greenwich Park; his old home in Bath, No. 19 New King Street, has been turned into a Herschel Museum. It was from the garden of No. 19 that Herschel discovered Uranus.

Herschel's main work was in connection with the stars, but he did make a series of observations of Mars between 1777 and 1783. He was fully convinced that the polar caps were due to thick masses of snow and ice, and he confirmed earlier suggestions that they are not centred exactly on the geographical poles. He measured the diameter of Mars, and fixed the rotation period at 24 hours 39 minutes 50 seconds. Later two German observers, Beer and Mädler, re-worked Herschel's observations and gave a revised period of 24 hours 37 minutes 23.7 seconds, which is only about one second too long.

Magnificent observer though he was, some of Herschel's ideas sound strange today; he believed that most worlds must be inhabited, and that there were even intelligent beings living below the surface of the Sun. To him, therefore, Mars was a world with lands and seas. He knew that there was an atmosphere, but he did not believe the atmosphere to be as dense as had been assumed by Cassini and others, and he proved this observationally. As it moves across the sky, Mars sometimes passes in front of a star, and hides or occults it; even if there is no actual occultation, a 'near miss' means that the star will pass behind the shell of atmosphere surrounding the planet, and will be noticeably dimmed. On 26 and 27 October 1783 Herschel watched two faint stars as they passed within a few seconds of arc of the Martian limb, and found that they did not fade at all. Presumably, then, the Martian atmosphere was not very extensive. Again Herschel was right. He also measured the flattening of the globe of Mars, which is more marked than that of the Earth, and he fixed the axial inclination at an angle of 28 degrees. As we have seen, the true value is 24 degrees.

Herschel's observations of Mars virtually ended in 1783, but two years later a long series was begun by Johann Hieronymus Schröter, chief magistrate of the little German town of Lilienthal, near Bremen. Schröter was in regular correspondence with Herschel, and one of his telescopes (probably his best one) was of Herschel's manufacture, but the two men were working along very different lines. Schröter was concerned entirely with the Solar System. He was the first really great observer of the Moon, and he paid careful attention to the planets, particularly Mars and Venus. He was president of an organization formed in 1800 to search for small planets moving round the Sun in between the orbits of Mars and Jupiter, and one of these 'asteroids' was discovered by his assistant, Karl Harding, working at the Lilienthal observatory.

I have always had a great admiration for Schröter. He has been accused (rightly) of being a clumsy draughtsman, and also (wrongly) of being an inaccurate observer. It has even been claimed that his telescopes were of poor quality, but this does not seem likely in view of the fact that one of them was made by Herschel. Unfortunately his observatory was destroyed in 1814 when the French invaded Germany, and the brass-tubed telescopes were plundered by the French soldiers under the impression that they were gold, but in all probability the Herschel telescope at least was of excellent quality. In any case, Schröter made a very useful series of drawings of Mars.

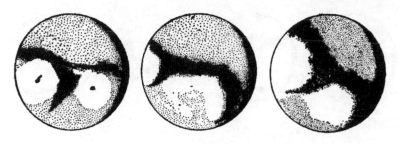

Drawings of Mars by Schröter, 8 December 1800.

He made various measurements (diameter, axial inclination, polar flattening and so on) which were better than Herschel's, but, surprisingly, he misinterpreted what he saw, and wrote that he had 'never observed with certainty any completely fixed dark patches which, like our seas and lakes, would have a lower reflecting power'. Surprisingly, he was convinced that the dark areas were simply clouds in the Martian atmosphere.

It is hard to see how he can have fallen into such a trap, particularly as he made many drawings which showed the same features time and time again. He also detected all the phenomena of the polar caps which are well known today, such as the variations in brightness and extent, the somewhat irregular outlines, and the seasonal cycle. It is a tragedy that so many of his observational books and notes were burned when the French sacked his observatory. This was in 1814; Schröter never recovered, and died two years later.

Before coming on to Beer and Mädler, who produced the first proper chart of Mars, I must say something about Honoré Flaugergues, a French amateur who is best remembered as the discoverer of the brilliant comet of 1811. Flaugergues set up his private observatory at Viviers, and observed Mars during several oppositions, including that of 1813. He was convinced that the markings were variable, but, unlike Schröter, he believed them to be true surface features. He also realized that during Martian spring and early summer a polar cap will shrink very quickly. Flaugergues assumed that the polar cap must be a thick layer of ice and snow, and he went on to draw a very remarkable conclusion: 'If the melting of the polar ices on Mars is much more prompt and much more complete than with our own terrestrial ice-caps, most of which persist up to the heat of summer, it seems therefore that the heat on Mars is greater than on Earth, though because of the planet's greater distance from the Sun it ought to be less in the ratio of 43 to 100. This is an extra reason to add to those which have made the most

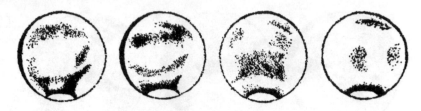

Drawings of Mars by Beer and Mädler, 1837.

skilful physicists believe that the rays of the Sun do not in themselves cause heat, but are only the indirect cause of heat.'

In other words, Mars has a climate hotter than that of the Earth! This sounds distinctly curious, but certainly there seemed no obvious reason why Mars should not support life.

Then, from 1830 to 1841, came the observations by Wilhelm Beer and Johann von Mädler, who worked together at Beer's private observatory outside Berlin. They made an interesting partnership. Beer – brother of Meyerbeer, the composer – was a banker by profession; Mädler was the main observer, and the telescope used was a modest 3¾-inch refractor. Despite this small aperture, Beer and Mädler compiled a map of the Moon which remained the best for half a century, and is still valuable today. They also drew up a map of Mars, and selected a dark feature to mark the zero for longitude – the Martian equivalent of Greenwich Observatory.* Relatively speaking, their map of Mars was not so good as their lunar chart, but it was certainly better than anything which had been produced before, and under the circumstances it was a remarkable achievement. Quite apart from this, they revised and improved all the earlier physical measurements – including the axial rotation period, which, as we have seen, they worked out to an accuracy of less than a second.

Mädler left Berlin in 1840–41 to become Director of the new Dorpat Observatory in Estonia, which was then under Russian rule. Neither he nor Beer did much more lunar or planetary work, which was a pity. It must be admitted that the following two decades were rather barren, apart from some good drawings made in the mid-1850s by a British amateur, Warren de la Rue, who was also a pioneer photographer. De la Rue used a 13-inch equatorial reflector, which must have been optically very good.

---

* Though Greenwich Observatory was not accepted as the zero for Earth longitude until much later (the 1880s).

The main feature on many of de la Rue's drawings, such as that of 20 April 1856, is the Syrtis Major, which extends across much of the disk and has a pronounced 'tail'. The southern polar cap is also well shown, and there are many other recognizable details. He also showed whiteness in the far north. In 1853 the French astronomer François Arago had already pointed out that the southern polar cap should show greater variations in area than the northern, because the climates there are more extreme.

In 1858 some detailed drawings of Mars were made by Angelo Secchi, a Jesuit who became famous for his pioneer work in the field of stellar spectroscopy. He recorded the Syrtis Major, and it was he who called it the 'Atlantic Canal', though the name could not have been worse chosen, and there was no suggestion that it might be artificial.

Up to that time it had been tacitly assumed that the dark patches on Mars were seas, while the ochre tracts were continents. One man who did not agree was Emmanuel Liais, who was trained in Paris but was then invited by the Emperor of Brazil to become Director of the Rio de Janeiro Observatory, where he spent most of the rest of his life. Liais made some rather desultory observations of Mars, but his most important contribution was the suggestion that the dark regions were likely to be vegetation-tracts rather than deserts. Liais' theory was published in 1860. Secchi did not agree, and two years later wrote that 'the existence of seas and continents . . . has been conclusively proved'. Even at that stage, Mars was showing itself to be a highly controversial world.

Better telescopes and better techniques led on to more accurate maps of Mars. Sir Norman Lockyer, another pioneer of stellar spectroscopy, produced some good drawings during the favourable opposition of 1862, and agreed with Secchi that the 'green' areas were oceanic. Frederik Kaiser in Holland and Camille Flammarion in France were other skilful observers; Kaiser gave the rotation period as 24 hours 37 minutes 22.62 seconds. Meanwhile, spectroscopy had been coming to the fore, and at first seemed to support the idea that the atmosphere of Mars was decidedly damp. Flammarion was certainly convinced that the dark regions were due to water in some form, though he did wonder whether it might be in a kind of intermediate state, neither pure liquid nor pure vapour.

Next, there was the problem of naming the Martian features. This was something which had never been seriously tackled, and the names used by different observers were unofficial. The lead was taken by Richard A. Proc-

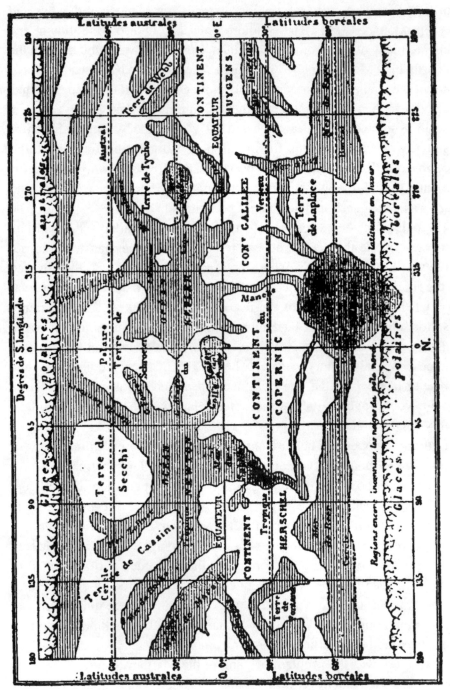

Map of Mars by Flammarion, 1876.

tor, a British amateur who was a noted 'popular' writer but was also a good astronomer in his own right. In 1867 Proctor produced a map in which he gave the Martian features names honouring famous observers – Cassini Land, Fontana Land, Mädler Land and so on. His system was followed by other British observers, but was less enthusiastically received on the Continent. Proctor's map was criticized as being inaccurate, and there was some truth in this; he was also attacked for having selected names which favoured Britons.* Various modifications were introduced, but the whole system was finally overthrown in favour of a new one by Giovanni Schiaparelli, introduced in 1877. I may be running ahead of my story, and if so I apologize; but it is interesting to compare the rival nomenclatures, and to look at two maps, one based on Proctor's system and the other on Schiaparelli's. The identifications are:

Proctor	*Schiaparelli*
Beer Continent	Aeria and Arabia
Herschel II Strait	Sinus Sabæus
Arago Strait	Margaritifer Sinus
Burton Bay	Mouth of the Indus canal
Mädler Continent	Chryse
Christie Bay	Auroræ Sinus
Terby Sea	Solis Lacus
Kepler Land and Copernicus Land	Thaumasia
Jacob Land	Noachis and Argyre I
Phillip Island	Deucalionis Regio
Hall Island	Protei Regio
Schiaparelli Sea	Mars Sirenum, Lacus Phœnicis
Maraldi Sea	Mare Cimmerium
Hooke Sea and Flammarion Sea	Mare Tyrrhenum and Syrtis Minor
Cassini Land and Dreyer Island	Ausonia and Iapygia
Lockyer Land	Hellas
Kaiser Sea (or the Hourglass Sea)	Syrtis Major

---

* Similar criticisms were levelled at the Russians nearly a century later, in 1959, when they sent their probe Lunik 3 round the Moon and obtained the first pictures of the far side, which we can never see from Earth because it is always turned away from us. Their system included, for example, the Moscow Sea and the Soviet Mountains. Alas, the Soviet Mountains do not exist; the feature shown was merely a bright ray, and the name was quietly dropped from the later Soviet charts!

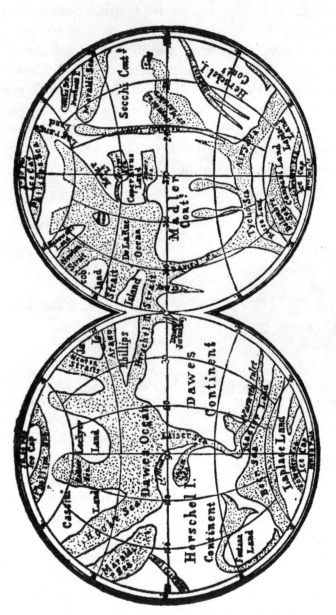

Map of Mars by Proctor, 1867.

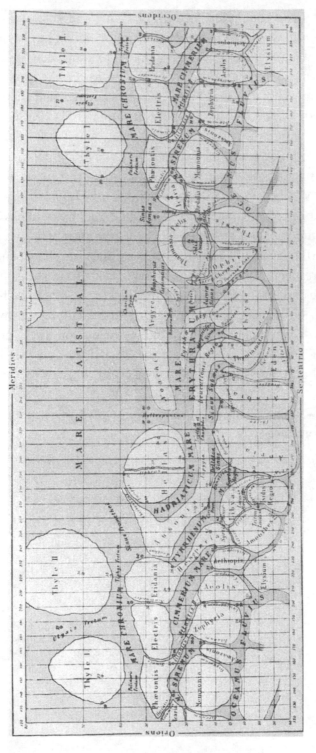

Map of Mars by Schiaparelli, 1877.

Which system do you prefer? I admit to having a feeling of nostalgia for Proctor's; after all, it is the kind of nomenclature which has been accepted for the Moon. On the other hand one is bound to admit that Schiaparelli's, based partly on geography and partly on mythology, is the more scientific. Yet Arago Strait is not a strait, Phillips Island is not an island, and the Hourglass Sea is not a sea – it is a lofty plateau sloping off to either side. Now that the space missions have told us so much more about Mars, Schiaparelli's plan has had to be revised; for instance his Lacus Phœnicis – the Phœnix Lake – has proved to be a towering volcano. Note that Chryse, the 'Golden Plain' where Viking 1 made its epic touchdown in July 1976, was the old Mädler Continent. One wonders what Johann Mädler would have thought.

Such was the situation at the end of what we may term the mediæval period of Martian history. The movements of Mars, its shape, and its main surface features were well known; the polar caps, varying seasonally, were assumed to be of ice or snow; the atmosphere was thought to be reasonably dense, though less so than that of the Earth, and the dark regions were either oceans (as most astronomers thought) or else vegetation-tracts, presumably filling old sea-beds. The temperature was expected to be only slightly lower than that of Earth, and the idea of advanced life there was certainly not ruled out.

The stage was set for one of the most memorable years in the history of Martian exploration: 1877.

CHAPTER 4

The Story of the Canals

To the early observers, Mars was an attractive world. It was also a mysterious one, because so little could be seen upon it apart from dark patches and white polar caps. Speculation was rife, and it is worth looking back to some comments by Bernard de Fontenelle, secretary of the French Academy of Sciences, in 1688. Fontenelle regretted that Mars had no satellites – at least so far as he knew; the two dwarf attendants were not discovered until almost two centuries later – and then went on as follows: 'We have seen phosphorescent materials, either liquid or dry, which upon receiving light from the Sun absorb it, so that they can shine brilliantly when in shadow. Perhaps Mars has great, high rocks, naturally phosphorescent, which during the day can store up light, emitting it again during the night. Nobody can imagine a pleasanter scene than that of rocks illuminating the whole landscape after sunset, and providing a magnificent light without inconvenient heat. In America we know that there are many birds which are so luminous that in darkness we can read by their light. How do we know that Mars does not have a great number of these birds which, when night comes, scatter on all sides and make a new day?'

This is at least far more inviting than the bloodthirsty Martians conjured up later by H. G. Wells and his imitators. In fact, the idea of hostile inhabitants did not occur to anyone until modern or near-modern times, and although men such as William Herschel believed firmly in the habitability of Mars they did not go so far as to speculate what the inhabitants might be like. Obviously the first requirement was to communicate with them, and various interesting suggestions were made. Johann von Littrow, who became Director of the Vienna Observatory in 1819, proposed lighting vast fires arranged in geometrical patterns to attract the attention of the Martians, who would presumably understand the message and make a suitable reply. Another bright idea was to dig wide trenches in the Sahara Desert, and provide a sort of mathematical code. The climax was reached in the mid-1870s by Charles Cros, an enthusiastic Frenchman, who put forward the scheme of building a large burning-glass which could focus the Sun's light and heat on to a Martian desert, scorching the sand there; by swinging the glass around it would be possible to write words on the surface of Mars. I have often wondered what words he intended to write, but the plan never progressed even as far as the drawing-board stage. Monsieur Cros was apparently most upset at the general refusal to take him seriously.

If we could not signal to the Martians, could we hope to see any evidence of their handiwork? This brings us straight on to the famous observations made in 1877 by Schiaparelli, setting off a violent argument which was not finally ended until the flights of the Mariners. It was Schiaparelli who made the first detailed studies of the features which he called *canali*, or channels, but which have been immortalized as the Martian canals.

Giovanni Virginio Schiaparelli was born in Piedmont in 1835, and graduated from Turin University. In 1862 he was appointed Director of the Brera Observatory in Milan, which was equipped with a fine 8¾-inch refractor. His interests were many (for instance, he carried out pioneer work in connection with comets and meteor streams), but for the moment we must confine ourselves to his work on Mars, which began with the 1877 opposition. The actual opposition date was 5 September, and Mars was practically at perihelion, so that it was to all intents and purposes as close to us as it can ever be. Schiaparelli was a skilled observer, and the Milan skies were clear (much clearer than they are today). He therefore decided to compile a new map of Mars.

The chart which he produced was certainly much better than any of its predecessors, and in the main it stands up quite well to the modern results. He also revised the nomenclature; out went Beer Continent, Lockyer Land and Dreyer Island, while in came Aeria, Hellas and Iapygia. For a while the two systems ran in parallel, but eventually Schiaparelli's prevailed, and it is indeed Schiaparelli's basic nomenclature that we use today, though the space mission revelations have meant that it has had to be somewhat modified.

The most striking features on Schiaparelli's maps were the very fine, regular lines running across the reddish-ochre deserts. They were, Schiaparelli believed, unlike anything else in the Solar System, and he was frankly taken aback. In a later article he summarized his ideas about them, so let us keep to his own words: 'All the vast extent of the continents is furrowed upon every side by a network of numerous lines or fine stripes of a more or less pronounced dark colour, whose aspect is very variable. They traverse the planet for long distances in regular lines, that do not at all resemble the winding courses of our streams. Some of the shorter ones do not reach three hundred miles; others extend for many thousands, occupying a quarter or even a third of the circumference of the planet. Some of these are very easy to see, especially the one designated by the name of Nilosyrtis. Others in turn are extremely difficult, and resemble the finest thread of a spider's web drawn across the disk. They are subject to great variations in breadth, which may reach 120 to 180 miles for the Nilosyrtis, while others are scarcely 20 miles broad . . . Their length and arrangement are constant, or vary only between very narrow limits . . . The canals may intersect among themselves at all possible angles, but by preference they converge toward the small spots to which we have given the name of lakes. For example, seven are seen to converge in Phœnicis Lacus, eight in Trivium Charontis, six in Lunæ Lacus and six in Ismenius Lacus.'

Perhaps the most remarkable thing about the canal network as described by Schiaparelli was that it seemed to follow a definite pattern. There was nothing haphazard about it. Either the canals followed great-circle tracks across the planet, such as the Phison, or else they were gently curved, such as the Nilosyrtis. Whether curved or not, they ran from dark area to dark area; there was not a single case of a canal breaking off abruptly in the middle of an ochre tract. Altogether, Schiaparelli recorded forty canals during the 1877 opposition.

It has often been said that Schiaparelli was not the first to see the canals, and that earlier observers had drawn some of them. This is basically correct. What seems to be a 'canal' is shown on one of the drawings made by Beer and Mädler, and there are streaks on the sketches made by observers such as Lassell, Lockyer, de la Rue, Secchi, Kaiser and others. The Rev. W. R. Dawes, in 1864, produced drawings on which there are streaks which would certainly have been called canals if the term had been invented then so far as Mars was concerned. But the aspect as shown by Schiaparelli was entirely different, and it opened up a new train of thought.

The next opposition was that of 1879. Mars was rather further from perihelion, but conditions were still good, and Schiaparelli made the most of them. He recovered the old canals, and added new ones. There was something more: single canals could be abruptly replaced by double ones, a phenomenon which became known as 'gemination' or twinning. To quote Schiaparelli again, when a gemination occurs, 'the two lines follow very nearly the original canal, and end in the place where it ended. One of these is often positioned as exactly upon the former line, but it also happens that the two lines may occupy opposite sides of the former canal, and be located upon entirely new ground. The distance between the two lines differs in different geminations, and varies from 370 miles and more down to the smallest limit at which the two lines may appear separated in large visual telescopes – less than an interval of 30 miles.' According to Schiaparelli's observations, a canal which appeared single one night might well be double the next.

When Schiaparelli first published his results there was a good deal of scepticism, which was understandable. Nobody else saw the canals in 1877, and although C. E. Burton in Ireland, using a 6-inch refractor and an 8-inch reflector, made a few sketches in 1879 which showed significant streaks, full confirmation of the canal network was not forthcoming for some time. Succeeding oppositions passed by with Schiaparelli still obtaining results which differed from those of other observers. For that matter, no two observers seemed to show Mars in the same guise.

Schiaparelli was not in the least deterred. Also, it had to be borne in mind that the oppositions of the early 1880s were less favourable than those of 1877 and 1879, because Mars was further from perihelion and its apparent diameter was smaller. Therefore the lack of confirmation was hardly unexpected, and Schiaparelli was convinced that the 'canali' were true channels, carrying away the flood-water produced by the melting of the ice-caps at the poles. At this stage he seems to have regarded them as natural, though afterwards he became more inclined to the view that they were artificial. He even wrote that 'Their singular aspect has led some to see in them the work of intelligent beings. I am very careful not to combat this suggestion, which contains nothing impossible.'

Finally, in 1886, came what appeared at that time to be full verification of the canal network. Using the powerful 30-inch refractor at the Nice Observatory in France, two professional astronomers – Perrotin and Thollon – published a chart which was as strange-looking as anything which Schiaparelli had produced. Subsequently, canals became thoroughly fashionable. It would be tedious to list all the observers who recorded

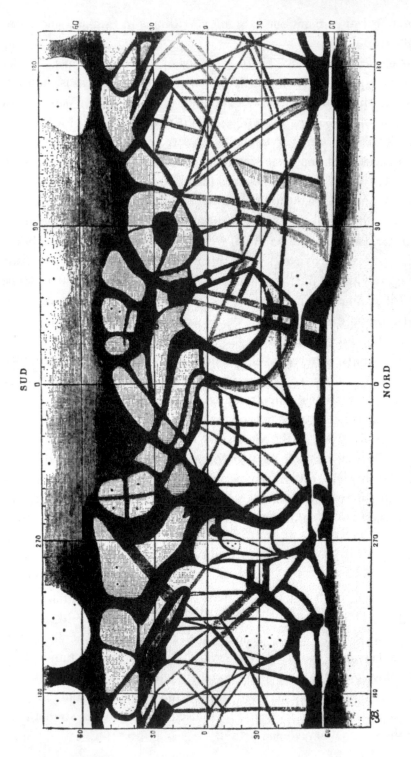

Martian canals as mapped by Schiaparelli, 1881–82.

them, though I must mention A. Stanley Williams in England and François Terby in Belgium.

Schiaparelli himself ended his main series of observations in 1890, because of eyesight trouble; he made some drawings afterwards, but was reluctant to publish them because he did not trust them. (Sad to say, he lost his sight completely a year or two before his death in 1910.) Meanwhile, two Americans had come very much to the fore: William H. Pickering and Percival Lowell. Pickering, a professional astronomer attached to Harvard, was an expert observer – he discovered the ninth satellite of Saturn, Phœbe – and was a lunar and planetary specialist. It is true that his ideas tended to be rather unconventional, and he tended to believe that some of the allegedly variable dark spots on the Moon were due to swarms of insects, but of his skill at the eye-end of a telescope there could be no doubt at all. He made numerous drawings of the Martian canals from 1892 onward, and it was he who gave the name of 'oases' to the dark patches in which the streaks intersected. He found that some of the oases were centres of radiating canals, and saw at least six issuing from the prominent dark area known as the Trivium Charontis, then regarded as a deep depression. More importantly, he discovered that there were canals crossing the dark regions as well as the bright deserts, which to all intents and purposes gave the death-blow to the idea that the dark areas were watery.

Yet the central figure in the Mars story became, and remained, Percival Lowell, who began his career as a diplomat and then decided to devote his life to astronomy. In 1894 he founded an observatory at Flagstaff, in Arizona, principally to study Mars; he equipped it with a fine 24-inch refractor, and he worked away untiringly until shortly before his sudden death in 1916.

Flagstaff was not chosen lightly. Lowell knew that as well as having a large telescope he had to have the clearest possible skies, and the field was wide open. Schiaparelli had used a relatively small instrument, but Lowell was well off, and he was able to go more or less where he liked. Tests showed that Flagstaff was exceptionally well-favoured climatically, and so to Flagstaff he went. The Lowell Observatory quickly became famous, and it still is. Nowadays there are many powerful telescopes there, and the 24-inch refractor has been relegated to a minor rôle, but it is undoubtedly one of the best instruments of its type. I have used it a great deal, so that I can speak from personal experience with it.

Lowell is remembered as the man who perfected the canal network, and who claimed that Mars must undoubtedly be inhabited by beings capable of building a planet-wide irrigation system, drawing water from the icy caps and transferring it to the populated regions closer to the equator. He was wrong on both counts. Yet we must also remember that he achieved a great deal of immensely valuable work in other fields, and it was he who made the calculations which led to the tracking-down of the ninth planet, Pluto; the discovery was made at Flagstaff, though admittedly not until years after Lowell's death. There can be no doubt that Lowell will always retain an honoured place in the history of astronomy.

During his lifetime he was joined, at various periods, by many skilled observers, including Pickering. Canals aplenty were recorded; geminations, variations, oases – in fact, phenomena of unique kind. Lowell was in no doubt about the reality of the canals. In his book *Mars and its Canals* (1906) he wrote that 'the Martian canals when well seen are not at the limit of visibility, but well within that boundary of doubt . . . under good atmospheric conditions the canals are comparable for conspicuousness to many of the well-recognized Fraunhofer lines, and are just as certainly there.' (The Fraunhofer lines are the dark lines seen in the spectrum of the Sun; many thousands have been mapped with absolute precision.)

Lowell's geometrical canal network could not, he maintained, be natural. Therefore the planet must support an advanced civilization, though he was careful to add that 'to talk of Martian beings is not to mean Martian men'. To him, the inhabitants had long since renounced warfare, which he described (accurately enough!) as 'a survival among us from savage times, and now affecting chiefly the boyish and unthinking element of the nation'. Mars was short of water because it was more advanced in its evolution than the Earth, and had aged more quickly; therefore, the Martians were doing their utmost to extract every scrap of moisture from the only reserves left – the ice and snow in the polar caps.

Two other phenomena seemed highly relevant. When a polar cap became smaller in spring and early summer, Lowell noted a dark band round its edge, which shrank as the cap shrank; this was attributed to moisture which produced temporary lakes or marshes, which would presumably be less reflective. Secondly, he claimed that the canals and the main dark areas were affected by the shrinking of the cap. According to his theory, the released moisture swept down toward the equator, and the vegetation revived from

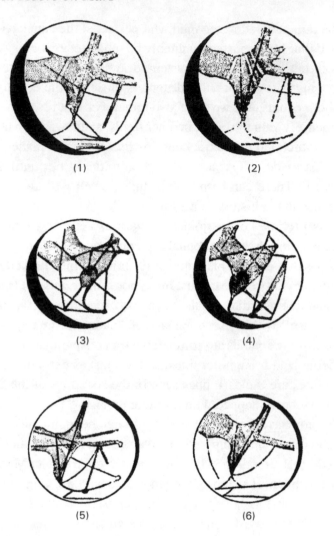

Drawings of the Syrtis Major region by Lowell: (1) 2 October 1896, 14.25; (2) 2 October 1896, 15.07; (3) 9 October 1896, 17.22; (4) 9 October 1896, 18.40; (5) 11 January 1897, 05.01; (6) 12 January 1897, 04.51.

its winter hibernation, so that a 'wave of darkening' was seen over a period of weeks or months. I will have more to say about this 'wave' later. Lowell even calculated that its rate was approximately two miles per hour. With regard to the dark areas, he wrote: 'Quickened by the water let loose on the melting of the polar cap, they rise rapidly to prominence, to stay so for some months, and then slowly proceed to die out again. Each in turn is thus affected . . . One after another each zone in order is reached and traversed, till even the equator is crossed, and the advance invades the territory of the other side. Following in its steps, afar, comes its slower wane. But already, from the other cap, has started an impulse of like character that sweeps reversely back again, travelling northward as the first went south. Twice each Martian year is the main body of the planet traversed by these waves of vegetal awakening, grandly oblivious to everything but their own advance. Two seasons of growth it therefore has, one coming from its Arctic, one from its Antarctic, zone, its equator standing curiously beholden semestrally to its poles.'

Lowell did not insist that a canal must be a channel of open water. Evaporation would be an obvious hazard, for instance. It seemed more likely that the narrow central canal was flanked to either side by strips of cultivated land, and the oases presumably represented population centres. But so far as Lowell was concerned, everything was dependent upon one claim: Mars was inhabited, and the canal network was artificial.

Not surprisingly, speculation followed speculation. Dykes, huge walls, and even strings of giant pipes were suggested. In a book called *The Riddle of Mars*, published in 1914, a writer named C. E. Housden went into great detail about the nature and arrangement of the pumping stations needed to transfer water from the poles to the equator, via the canals. Housden proved to his entire satisfaction that the water-bearing pipes must be about six feet in diameter, and were used to force the water up to high-level service reservoirs by way of the canals, from whence the water was distributed to the lands by static pressure. Coming on to more modern times, Donald Lee Cyr, in 1944, attributed the canals to fertility tracts, made by bands of creatures tracking across the surface from one oasis to another – though Cyr believed the oases to be natural craters on Mars produced by the impacts of meteorites. Cyr seems therefore to have been the first to publish a suggestion that there might be craters on Mars. He was followed by E. J. Öpik some time later, but Öpik's ideas were very different from Cyr's.

Not everyone believed that the canals were artificial. Pickering considered that they were natural cracks in the surface, through which gases could escape from the interior and which were filled with vegetation. And even in Lowell's lifetime there was strong criticism of his theories, which were regarded as extreme. One violent attack was launched by Alfred Russel Wallace, one of the pioneers of the theory of evolution, who wrote a whole book on the subject in 1907, concluding that water must be absent from the planet and that the temperature was hopelessly low. He ended with the sentence: 'Mars, therefore, is not only uninhabited by intelligent beings such as Mr. Lowell postulates, but is absolutely UNINHABITABLE!' (The capitals are Wallace's.)

If Lowell's maps had been accurate, there could have been little doubt that the features were non-natural. But the crux of the whole problem was – Did the canals exist at all? Were they straight and regular; were they broader, diffuse streaks totally unlike artificial waterways; or were they nothing more than tricks of the eye?

Significantly, the maps of the network drawn by different observers did not agree at all well. Apart from the most obvious of the streaks, the 'canals' shown by Schiaparelli, Lowell, Perrotin and Thollon, Leo Brenner and others were in different positions and of different intensities. Moreover, in some charts – notably Brenner's – the complexity of the system was unbelievable by any standards. This is not to say that there was no agreement at all; there were a few canals, notably the Nilosyrtis and the Nepenthes-Thoth, which were shown by almost everyone who could see the network at all. Yet even here, some observers showed them as broad, diffuse stripes, while others preferred the spider's-web appearance drawn by Lowell and his colleagues.

N. E. Green, the pioneer English observer who had been making careful studies of Mars well before the revelations of 1877, suggested that the canals were not true streaks, but merely the boundaries between areas of different colours. Vincenzio Cerulli, an Italian enthusiast who established an observatory at his home town of Teramo, proposed that the canals were not straight and regular lines, but were made up of disconnected spots and streaks; as he pointed out, the human eye does tend to join small features together when straining to glimpse objects at the very limit of visibility. Cerulli was unprejudiced, because he could himself see what might be interpreted as canals; he had no faith in Lowell's interpretations, but neither did he dismiss the canal system as a pure illusion.

Cerulli's theory was tested in 1903 by E. W. Maunder, in England. He made some drawings of Mars, without canals, and showed them at a distance to a class of boys from the Royal Greenwich Hospital School, telling them to make copies. When the boys did so, many of them showed sharp, linear canals, and Maunder drew the obvious conclusion. Lowell was not impressed. He dismissed the idea contemptuously as the 'small boy theory'.

For my own satisfaction, I repeated the experiment in 1950. The boys, at a Kent preparatory school, were between ten and thirteen years old – rather younger than those in Maunder's class, but of a higher educational level, and more used to drawing. The pictures shown were actual drawings of Mars made through a large telescope, but with disconnected spots and streaks put in along the alleged canal sites. Out of a total of 58 boys, 42 showed vague indications of 'something' where the canals should have been, 13 showed continuous broad, hazy strips, and only 3 showed Lowell-type canals. Of these three, two of the boys were notoriously unartistic and the third short-sighted. I give the results here to demonstrate that Lowell did, frankly, have right on his side when he preferred to trust trained observers rather than schoolboys!

And yet there were experienced observers who completely failed to see the canals in any form whatsoever. One of these was George Ellery Hale, planner of the great reflectors at Mount Wilson and Palomar. Another was Asaph Hall, who had discovered the two Martian satellites in 1877. Most astronomers believed Lowell to be wrong in claiming that the network was glaringly obvious, and after Lowell's death in 1916 the doubters grew in number.

Probably the most skilful planetary observer of the inter-war years was Eugenios M. Antoniadi, a Greek-born astronomer who went to France and spent most of his life there. For a time he worked with Camille Flammarion, the great popularizer of astronomy (and, let it be added, an excellent ob-server in his own right). Later Antoniadi was able to make extensive use of the 33-inch refractor at Meudon, near Paris, which again I know well and which is a superb instrument. With it, Antoniadi drew up the best map of Mars to be compiled before the space-probe era. He could see 'canals', using the term in its most general sense, but to him they were by no means artificial in appearance, and he had no patience whatsoever with Lowell's work, either observational or theoretical. Finally, in 1930, he published a monumental work which contained a complete description of the Martian surface.*

Antoniadi's views were clear-cut. 'Here is the actual truth of the matter. Nobody has ever seen a genuine canal on Mars, and the more or less recti-

---

* For some reason which I cannot explain, it was not translated into English for many years. Eventually I produced a translation, which was published – but not until 1975, by which time it was of historical interest only.

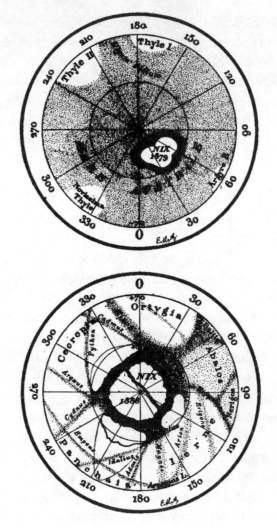

Polar regions as shown by Flammarion and Antoniadi.

linear, single or double "canals" of Schiaparelli do not exist as canals or as geometrical patterns; but they have a basis of reality, because on the sites of each of them the surface of the planet shows an irregular streak, more or less continuous or spotted, or else a broken, greyish border or an isolated, complex lake . . .   It is necessary to add here that Schiaparelli's "canals", which have a basis of reality, are quite different from the completely illusory "canals" shown on the drawings of Mars . . . by Lowell and his assistants. Several authors have compared Lowell's charts of Mars sarcastically with spider's webs. Such a comparison is not inappropriate.'

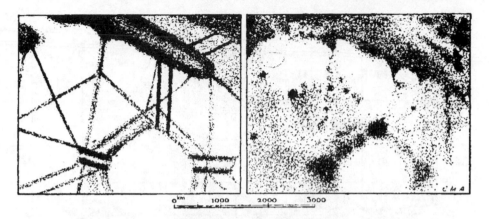

Antoniadi's explanation of the canals. (Left) Canals in the region of Elysium, drawn by Schiaparelli between 1877 and 1890. (Right) The same region, drawn by Antoniadi between 1909 and 1926 with the Meudon 33-inch refractor; the hard, linear features are broken down into fine detail.

We now know that Antoniadi was correct in rejecting the Lowell network, but in my view he was being a little harsh, because it is now clear that Schiaparelli's canals do not exist either. Antoniadi did not hesitate to point out that Lowell had also drawn linear features upon other planets, particularly Mercury and Venus, which are equally non-existent. So far as the oases were concerned, Antoniadi rejected them because he had seen them 'perfectly round even when a long way from the centre of the disk'; had they been real, they would have been foreshortened.

Antoniadi's opinions carried a great deal of weight. (He died in Occupied France during the Second World War.) But even in the years when rocket probes were being developed, the canal puzzle was still a matter of debate. Users of small telescopes often produced drawings which were very much on Lowell's pattern, and as recently as 1959 Clyde Tombaugh, who discovered the planet Pluto in 1930 while working at Flagstaff, proposed that the oases were impact craters, while the canals linking them were natural cracks filled with vegetation – not an original idea, since something of the sort had been suggested long before by the Swedish scientist Svante Arrhenius, but one which was still regarded as credible.

During the 1950s, a long series of observations was carried out by Audouin Dollfus at the Pic du Midi Observatory in the French Pyrenees. The refractor there, a 24-inch, is as large as Lowell's, and the Pic is 10,000 feet high, so that the atmosphere is exceptionally steady and transparent. Broadly,

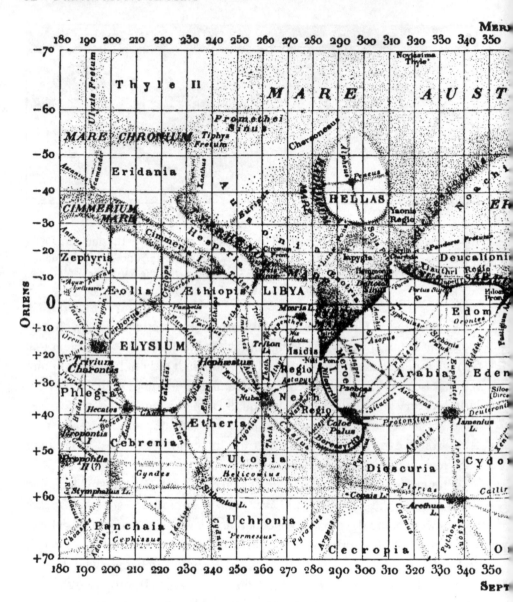

Part of Mars on Mercator's Projection by Antoniadi. Prepared from
the observations of the BAA Mars Section in 1903 and reproduced by kind permission
of the British Astronomical Association.

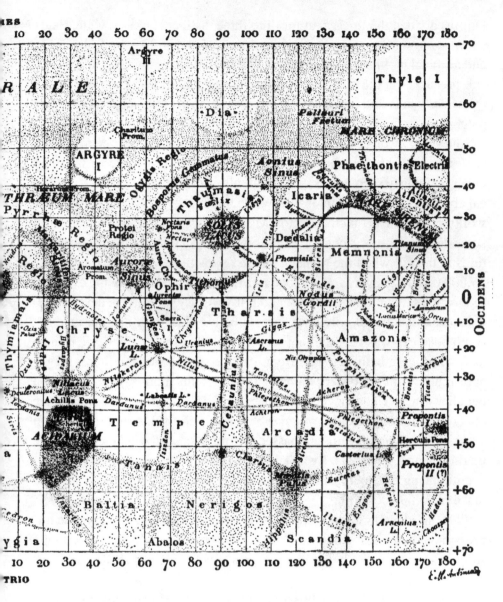

M. = Mare; S. = Sinus; Fr. = Fretum; L. = Lacus; P. = Palus; F. = Fons;
R.= Regio; I. = Insula; Pr. = Promontorium.

Dollfus divided the linear features into three distinct classes – wide, shady bands; narrow, more regular streaks; and 'canals', thread-like, perfectly black, and artificial looking. He maintained that the latter were purely physiological, produced by the human eye, while under conditions of perfect seeing the wider bands and streaks could be broken down into separate dots and patches.

And yet . . . Consider the observation by R. S. Richardson, one of America's most respected astronomers, made on 4 June 1956. Richardson had never previously seen canals, despite all his efforts, and was frankly sceptical about them. His account is well worth quoting: 'I was on Mount Wilson working with the 60-inch reflector. My object was to take some test exposures on Mars with various filters and emulsions in preparation for observations in the fall when the planet came to opposition. Mars at the time was 75,000,000 miles away . . . As soon as Mars came into view in the eyepiece I knew the seeing must be pretty good. It was easy to hold for several seconds detail that could only be suspected before. The dark band around the south polar cap was very evident . . . A magnification of 700 made it seem as if Mars were being viewed from a space-ship. I was familiar with the appearance from my observations of 1954, but this morning the disk had a peculiar aspect I had never noticed before. The bright red regions were covered with innumerable irregular blue lines, like veins running through some mineral. Some minutes passed before it occurred to me that these markings must be canals. I was taken completely by surprise, as I had not been thinking of canals, and certainly did not expect to see them at a distance of 75,000,000 miles.'

True, he added that 'their appearance was not artificial, but gave the impression of being some natural feature'. Observations of this sort certainly supported Antoniadi's belief that the canals had a basis of reality – at least in some cases.

At this stage I think it may be worth saying something about my own observations; after all, I have been looking at Mars for well over sixty years now. Before the war I used small telescopes of the 3-inch to 12-inch class, and of course I saw no canals. (It is true that many sketches made with apertures of this kind have shown canals in plenty, but although I have no doubt of the honesty of these observers it is obvious that they were unconsciously drawing upon their imagination; it is only too easy to 'see' what one half-expects to see.) After the war ended, and I emerged from the Royal Air Force, I became officially in-

I made this sketch in 1973 using a 27-inch refractor (0208–0223
G.M.T., magnification 580) and found no evidence of the canal network.

volved in mapping the Moon, so that I had access to large telescopes – mainly
refractors, which are ideal for this kind of work.

The first great telescope which I used systematically was the 33-inch
refractor at Meudon, outside Paris, with which Antoniadi had done most of
his work. I saw no canals. Then I was able to go to the Lowell Observatory
at Flagstaff, to use the 24-inch refractor. I will always remember my feeling
of excitement when I had my first view of Mars through Lowell's telescope:
would I see his canals? No. I did my best, but there was no way in which I
could see any trace of the canal network, though fine details on the disk
showed up magnificently. Later I was equally unsuccessful with the Innes
27-inch refractor at Johannesburg in South Africa, and even with the 60-
inch reflector at Palomar Observatory in California. To me, the canals simply were not there.

The coup de grâce to the canal network was given by the Mariner space-
craft, between 1965 and 1972. They showed craters, volcanoes, ravines,
canyons and much else, but of the canals there was no trace, and finally
even the most ardent Lowell supporters had to admit that the canals were
due to nothing more than tricks of the eye.

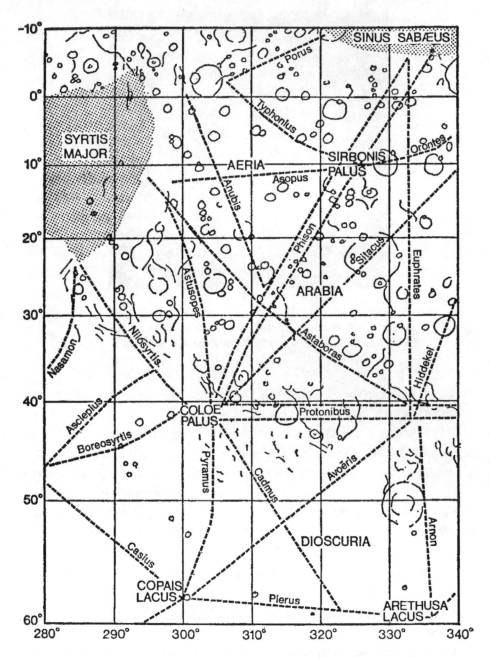

The region of Aeria, Arabia and Dioscuria. The details are taken from the maps obtained from the Mariners and Vikings, and the canal network is superimposed. It is clear that there is no correlation whatsoever and that the canals do not correspond to any real features.

And yet – could there be any correlation at all with real surface features, as Antoniadi had believed? I was not sure, so I decided to make a practical test. I selected part of one of the modern maps of Mars, obtained from the Mariner 9 pictures, and superimposed upon it the canals as they had been shown in a map published in 1910 by Flammarion and Antoniadi. I chose this region because it includes the Syrtis Major and the Sinus Sabæus, as well as some of the canals whose existence was regarded as absolutely definite – the double Phison, the Euphrates, the Arnon, the Hiddekel, the Casius and so on. There was absolutely no correlation, so that none of these canals can be said to have even a 'basis of reality', to use Antoniadi's own term. There are vague darkish patches in the sites of the old Ismenius Lacus and Coloe Palus, but no sign of any radiating streaks, while another reported dark patch, the Siloe Fons, seems to be completely non-existent.

The canals have their place in history. Most people – including myself – regret that they have been disproved. But Science is remorseless, and the long argument is over. Nobody has ever seen a true canal on Mars; there are no canals to see.

CHAPTER 5

Mars Before the Space Age

It is fair to say that the story of our exploration of Mars can be divided into several periods, each of which is reasonably well-defined. They are: (1) Very early days, pre-1830, when only the main features were known – that is to say, the dark areas, the ochre tracts and the white polar caps. (2) 1830 to 1877, when maps became at least reasonably accurate, and telescopes were powerful enough to show Mars in considerable detail. (3) 1877 to 1909, when canals were very much in vogue and many astronomers followed Lowell in believing that there might well be intelligent life there. (4) 1909 to 1965, when canals and Martians were generally rejected, and more modern-type techniques provided information which was thought to be reliable. (5) 1965 to the present time, when unmanned space-probes have at last enabled us to find out what the planet is really like.

The final period began with the first successful Mars mission, Mariner 4. In July 1965 it passed Mars at a range of 6200 miles, and sent back over twenty images obtained at close range as well as providing data which showed that the planet's atmosphere was much thinner than had been expected. Up to that time almost everyone had been fairly sure that the dark areas were tracts of vegetation, but within a few years after the first encounter this idea too had to be cast on to the scientific scrap-heap. All in all, Mars was not the sort of world which it had always been thought to be. So let us look back to the pre-Space Age days, and see how our opinions have changed. Some of the older ideas were right; others were most definitely wrong.

First, the atmosphere. Of its existence there had been no serious doubt since the eighteenth century. As long ago as 1785 William Herschel had written that 'Mars is not without considerable atmosphere; for besides the permanent spots on its surface, I have often noticed occasional changes of partial bright belts; and also once a darkish one, in a pretty high lati-tude. And these alterations we can hardly ascribe to any other cause than the variable disposition of clouds and vapours floating in the atmosphere of the planet . . . [There is] a considerable but moderate atmosphere, so that its inhabitants probably enjoy a situation in many respects similar to our own.'

As we have noted, a thin atmosphere on Mars would be expected from the rather low escape velocity of 3.1 miles per second, and over the years many attempts were made to measure the ground pressure. Unfortunately all the methods used were subject to very considerable doubt, and before the Space Age there was no way round the problem. One line of research was to see how the brightness of various patches on the disk varied according to the distance from the central meridian, since a feature near the limb would be seen through a thicker layer of Martian atmosphere than one near the centre of the disk. In 1939 Gérard de Vaucouleurs, at Le Houga Observatory in France, made a long series of estimates, and came to the conclusion that the ground pressure must be about 7 centimetres of mercury, or slightly more than 90 millibars, as against an average pressure of 960 millibars at sea-level on the Earth. Various other methods were also used. Since all are now obsolete there is no point in going into details about them, but it is interesting to list some of the values derived:

1934	N. Barabaschev (U.S.S.R.)	50 millibars
1940	N. Barabaschev (U.S.S.R.)	116 ,,
1941	V. V. Sharonov (U.S.S.R.)	120 ,,
1944	N. Sytinskaya (U.S.S.R.)	112 ,,
1945	G. de Vaucouleurs (France)	93 ,,
1948	S. L. Hess (U.S.A.)	80 ,,
1951	A. Dollfus (France)	83 ,,
1960	E. Öpik (Ireland)	116 ,,

Apart from Barabaschev's early estimate, there seemed to be general consensus of opinion that the ground pressure of the Martian atmosphere was between 80 and 120 millibars – about the same as that in the Earth's air at a height of eleven miles above sea-level. Remember that Mount Everest is less than six miles high, so that even if the Martian atmosphere had been made up of pure oxygen it would still have been of little use to beings such as ourselves. Incidentally, it was obvious that a terrestrial barometer would have to be re-calibrated for use on Mars.

There was one very important conclusion to be drawn. With a ground pressure of around 85 millibars, liquid water could exist on Mars provided that its temperature did not rise much above 100 degrees Fahrenheit. Of course, few people believed that water would be found there; all the same, it was not regarded as theoretically impossible. Moreover, the lower gravity would mean that the Martian atmosphere would fall off in density, with increasing height, much more gradually than on the Earth. At 18 miles up, the pressures would be the same, and at greater altitudes the atmosphere of Mars would actually be the denser of the two.

Then, however, new techniques came into play, and there were indications that the atmospheric pressure was much lower than 80 millibars. The Mariner findings were conclusive. On Mars, the pressure of the atmosphere was shown to be below 10 millibars everywhere, and when the Vikings landed, in 1976, they reported pressures of below 7 millibars. The pre-1960 estimates had been wrong by a factor of more than ten.

Estimates of the atmospheric composition had been even worse. Pre-Mariner, the only possible method was to use the spectroscope, which splits up light. Mars shines by reflecting the rays of the Sun, and basically its spectrum is simply a much enfeebled version of the solar spectrum; but a ray of light reaching us from the planet has passed through the Martian atmosphere twice – once on its way from the Sun to Mars, and once on its way back from Mars to Earth. Oxygen and water vapour, if present, would be expected to leave definite imprints upon the spectrum we see, and this also applies to other gases, though nitrogen – which makes up about 78 per cent of the Earth's air – is very shy about revealing itself in the visible spectrum under such conditions.

The first serious measurements were encouraging. In 1867 Jules Janssen, who founded the Meudon Observatory (the square outside the main entrance is still known as the Place Janssen) took his instruments up to the top of Mount Etna, 9800 feet above sea-level, to study the spectrum of Mars without the handicap of having to look through the denser lower parts of our own atmosphere, which complicates matters because it contains a great deal of both oxygen and water vapour. Janssen's results seemed to indicate that the Martian atmosphere was fairly rich in water vapour, at least. Studies by the great English spectroscopic pioneer, Sir William Huggins, and by Hermann Vogel in Germany provided confirmation – or so it was thought.

The essence of Janssen's method was to compare the spectrum of Mars with that of the Moon. Both shine by reflected sunlight, but the Moon has no atmosphere at all, so that any differences in the two spectra would presumably be due to gases around Mars. Janssen's reasoning was perfectly sound, but his equipment was very crude judged by modern standards, and with hindsight we have to admit that his results were illusory.

The next major step was taken by W. S. Adams and T. Dunham in 1933, at the Mount Wilson Observatory in California. Their instruments were far more sensitive than Janssen's, and they followed a different line of attack. The positions of the lines in a spectrum will be affected by the motion of the body sending out the light; this is the well-known Doppler Effect. If the light-emitting body is approaching, the spectral lines will be shifted over toward the short-wave or 'blue' end of the spectrum; if the body is receding, the shift will be to the 'red' end. When Mars is coming toward us, therefore, the lines due to (say) oxygen in the Martian atmosphere will be blue-shifted, whereas the lines due to the oxygen in our own air will be unaffected. In this way it ought, theoretically, to be possible to disentangle the two sets of lines, and so to work out how much oxygen there is around Mars.

The results were interesting. Adams and Dunham failed to find any oxygen at all, and wrote that 'the amount of oxygen in the atmosphere of Mars is probably less than one-tenth of 1 per cent of that in the Earth's atmosphere over equal areas of surface'. Later investigators were equally unsuccessful. In 1947 G. P. Kuiper, also in the United States, announced that he had detected traces of carbon dioxide; in 1963 Dollfus, working at the Jungfraujoch in the Swiss Alps, claimed that there was a certain amount of water vapour, but it was tacitly assumed that the bulk of the Martian atmosphere must be made up of nitrogen. This would have been no surprise. By volume, our own air is made up of about 78 per cent nitrogen, 21 per cent oxygen, and 1 per cent of other gases, mainly argon. Carbon dioxide accounts for a mere 0.03 per cent.

Historically, I must not omit to mention a theory proposed in 1960 by three American astronomers: C. C. Kiess, S. Karrer and H. K. Kiess. They believed that the atmosphere of Mars contained poisonous oxides of nitrogen, and that the polar caps were due to deposits of solid nitrogen tetroxide, while the reddish colour of the planet was caused by nitrogen peroxide. Not many people agreed with this idea, and if it had been correct our hopes of establishing colonies on Mars would have been permanently dashed. Luckily, we now know it to be wrong.

For that matter, the best 'official' analysis of the Martian atmosphere in pre-Mariner days was as wide of the mark as it could possibly be. De Vaucouleurs summed matters up when he listed the probable composition as 98.5 per cent nitrogen, 1.2 per cent argon, 0.25 per cent carbon dioxide and less than 0.1 per cent oxygen.

If this had been correct, and the surface pressure had been in the region of 87 millibars, Mars would have seemed to be a world which could well support low-type life at least. Unfortunately the space-probe analyses showed it to be completely wrong. Carbon dioxide makes up 95.3 per cent of the atmosphere; nitrogen accounts for 2.7 per cent and argon for 1.6 per cent, leaving less than one per cent for everything else. From our point of view, then, the atmosphere is more or less useless. It is very poor at retaining the Sun's daytime warmth (modest though this is), so that the nights are very cold indeed; even at the equator in midsummer the temperature falls well below anything we encounter on the surface of the Earth, even at the South Pole. There is virtually no protection against larger bodies, and as a shield against short-wave radiations the atmosphere is much less effective than ours. Yet despite the very small amount of water vapour, the night-time atmosphere is close to saturation point. Winds may be rapid, but in that tenuous atmosphere they will have very little force, so that a Martian 'gale' will be very modest by terrestrial standards.

Another problem concerned the so-called 'violet layer' – and although we now know that it does not exist, the story is interesting enough to be described in some detail.

Light may be regarded as a wave-motion, and the colour depends on the wavelength. In the visible band, red has the longest wavelength and violet the shortest, with orange, yellow, green and blue in between; anyone who has seen a rainbow will be familiar with this order. Considering its comparative thinness, the Martian atmosphere is strangely opaque to short wavelengths. Under normal conditions, a photograph of the planet taken in violet or blue light will appear blurred and slightly enlarged, because the rays do not penetrate to the surface at all, and all that we are photographing is the upper part of Mars' atmosphere. Photographs taken in red light slice through the shielding layers and record surface details, such as the dark areas and the polar caps, with no difficulty at all.

So far as the short wavelengths were concerned, it was claimed that something acted as a screen, and that this 'something' was variable. There were times when it cleared away, but it always came back. This might indicate a definite layer of material which was practically opaque to short-wave light, and this became known as the Violet Layer, not because it looked violet – visually it could not be seen at all – but because it blocked out the violet and blue rays.

The Layer was first reported in 1909, by C. Lampland at Flagstaff. Until 1937 it was assumed to be permanent, but in that year E. C. Slipher, one of the most famous of all observers of Mars, found that at intervals the atmosphere clears 'sufficiently to permit the short wavelengths to penetrate to the surface below and come out again'. Using the equipment at Flagstaff, Slipher took many hundreds of photographs, and maintained that during a 'clearing' of the Layer, photographs taken with blue or violet filters showed almost as much detail as the red and yellow pictures normally do.

Sometimes the clearing is partial; sometimes it affects the whole of Mars. Major clearings were seen in May 1937, July 1939, October 1941, December 1943, and during the oppositions of 1954 and 1965. In the latter year, according to de Vaucouleurs, it lasted from 25 August to 3 September. There was a partial clearing in 1958, and others in 1964 and 1965.

All sorts of theories were proposed to explain the Layer. One, widely favoured, was that it was due to tiny ice crystals from 0.3 to 0.4 microns in diameter (one micron is equal to 1/10,000 of a centimetre). This was the view held by Kuiper and Hess, both of whom had recorded clearings. Ice crystals, they wrote, could account for the Layer, and a rise in temperature of only a few degrees would be sufficient to make the crystals 'sublime' – that is to say, change directly from the solid to the gaseous state without passing through a liquid stage. This would at least explain the suddenness of the clearings, but there were various major objections. For instance, in 1963 de Vaucouleurs studied Mars in the ultra-violet region of its spectrum. Ice crystals reflect strongly in the ultra-violet, but Mars does not.

It was also pointed out that if the clearings were due to the sudden disappearance of the ice crystals, one would have to suppose that the temperature rose abruptly and uniformly all over Mars, which did not sound reasonable. The same objection was raised against the idea that the crystals were composed of solid carbon dioxide – which would, in any case, make the Layer more opaque than it was actually observed to be.

In 1960 E. J. Öpik revived an idea that the Layer could be due to small particles of carbon black, and he suggested that the absorbing material was made up partly of these carbon black particles and partly of dust. Clearings would occur when the dust settled down during calm periods. Unfortunately for this theory, the major clearing of 1956 took place at the time of a great dust-storm. Still another theory was due to the Czech astronomer F. Link, who believed that the Layer particles were interplanetary dust which had collected in the upper Martian atmosphere. (*En passant*, it is certain that meteors could be seen from Mars, because the atmosphere, thin though it is, is quite sufficient to act as a shield. Meteors which dash into our own air, and burn away by friction against the air-particles, are usually destroyed at a height of more than 40 miles above the ground, and at this level the atmospheric density is considerably less than it is at the surface of Mars.)

One fascinating study was carried out by S. L. Hess in 1941, when there was a clearing which persisted for several days. The Layer, he said, was normally able to protect the surface of Mars from short-wave radiations, just as the ozone in our own stratosphere protects the Earth and makes life here possible (though let me stress that nobody seriously believed the Violet Layer over Mars to be due to ozone). During a clearing, Mars is exposed to the full short-wave bombardment, which might be expected to damage any living organisms. At that time it was still confidently believed that the dark regions such as the Syrtis Major were due to vegetation of some kind, and that this vegetation could develop only when it could take in sufficient moisture. It was claimed that in the Martian spring, when the polar cap began to melt, water vapour was released and wafted down from the pole toward the equator, so that there was a 'wave of darkening' as the vegetation started to revive after the arid winter. Hess found that in 1941, when there was a prolonged 'clearing' due to the absence of the Violet Layer, the development cycle stopped, to recommence only when the Layer reformed and conditions reverted to normal. This appeared to confirm not only the existence of a definite screen, but also that the dark areas were due to something which lives and grows.

It sounded very convincing, but it too turned out to be wrong. There is no Violet Layer. The 'clearings' are due to phase effects of the scattering of light by atmospheric dust, reducing the contrast differences between the bright and darker areas when viewed in light of short wavelength. But what about the 'wave of darkening'?

It used to be regarded as a definite phenomenon, and was well described by Gérard de Vaucouleurs, one of the leading French planetary scientists, from his observations made at the favourable opposition of 1939, when the south pole of Mars was turned in our direction. 'The greater the distance of the dark areas from the south polar cap, the later the darkening begins; we see, in short, a wave of darkening proceeding from the south polar regions. A wave which begins at the end of the winter at the border of the cap, near latitude 60 degrees south, reaches the equator by the middle of spring, passes into the northern hemisphere and extends, before the end of the southern spring, as far as latitude 40 degrees north, travelling thus some 3800 miles in about 130 days, or about 29 miles per day.' This was certainly precise, but the agreement between different observers was by no means complete, and doubts began to creep in.

When I was able to use powerful telescopes, after the end of the war, I did my best to follow the wave of darkening. I failed completely. Of course there were minor variations in the colours and occasionally the shapes of the dark areas, but nothing at all systematic, and by now it is generally agreed that the wave of darkening is as unreal as the Violet Layer or, for that matter, the canals.

Clouds, of course, have been under observation for a long time; so far as I know they were first mentioned by Honoré Flaugergues as far back as 1809. They used to be divided into two classes, white and yellow, and this is still true, though the yellow clouds are now officially regarded as dust-storms.

One particularly interesting feature on Mars is the circular Hellas, south of the Syrtis Major; it was called Lockyer Land on the old maps. Sometimes it is so brilliant that it can easily be mistaken for an extra polar cap, though at other times it is dull. It was once believed to be a lofty plateau which could be covered with snow. Today we know it to be a basin – in fact, the deepest on Mars – and it can sometimes be cloud-filled, which accounts for its brilliance. Localized white clouds may be seen anywhere, and so too may the sunrise and sunset fogs and hazes. There are also ice crystal clouds high in the Martian atmosphere. These are never conspicuous as seen from Earth, but in 1998 excellent pictures of them were taken from the Pathfinder probe resting on the planet's surface.

'Yellow' clouds can also be localized, but we also see major dust-storms which spread out until they may cover the whole of the planet, hiding the familiar surface features. Generally they start at high latitudes, and they are commonest when Mars is near perihelion and the surface winds are at their strongest. What seems to happen is that if the wind-speed exceeds a certain critical value (150 to 300 feet per second) grains of surface material are whipped up and given a kind of skipping motion, known technically as saltation. On striking the surface they propel even smaller grains into the atmosphere, where they remain suspended for weeks – or, in extreme cases, months.

Major dust-storms were recorded at the perihelic oppositions of 1909, 1924, 1929, 1941, 1956 and 1971–72. That of 1909 was particularly intense, and was carefully followed by E. M. Antoniadi, who had the advantage of being able to make use of the 33-inch refractor at the Meudon Observatory. He was also observing in 1924, and during August reported that 'the planet had become covered with yellow clouds, and presented a cream colour similar to that of Jupiter'. I was unable to follow that storm personally, as I was then at the early age of one, but I did see the storms of 1956 and 1971–72. I made a long series of observations in 1956, when Mars was almost as close to us as it can ever be. Using my 12½-inch reflector, I was able to see

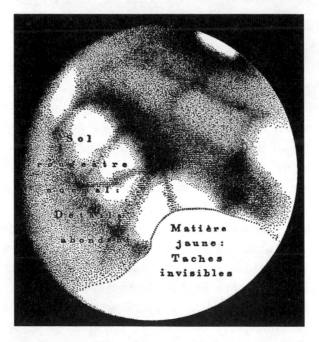

Vast yellow cloud masking the Trivium Charontis and neighbouring regions, observed by Antoniadi from 23 to 27 August 1929 (8½-inch reflector).

the usual surface features throughout August and the first part of September. (Opposition fell on 10 September.) Yet as early as 2 September it became plain that something was amiss. The polar cap was becoming obscure; I could still see it on 8 September, but at that point I decided that my reflector was not large enough to satisfy me, so I took advantage of an invitation to Meudon. From there, on 12 September, I observed Mars under excellent conditions. The disk was practically blank, and remained so for more than a week. The cap reappeared about 19 September, but it was mid-October before the dust-storm finally cleared away.

The next planet-wide storm was that of 1971. It began in mid-September, and was first photographed by Gregory Roberts, using the 27-inch refractor at Johannesburg. Observers at the Lowell Observatory at Flagstaff reported that it spread out from an initial streak-like core around 1500 miles long, and expanded until by 27 September it covered a large area touching the eastern edge of the famous circular feature Hellas. I could still see Hellas at that time, but it had lost its whitish hue and was no more than very slightly brighter than its surroundings. On 28 September I recorded well-marked dark features in their familiar guise; then came two cloudy nights from my Selsey observatory, and when I could observe again, on 1 October, there was a striking transformation. All I could see were vague indications of the south polar cap and a few indefinite shadings. Within a week I had lost even these, and it was not until the end of December that anything positive could be made out once more – by which time Mars had receded so far from the Earth that details were by no means easy to see even without the added disadvantage of Martian dust.

This particular storm came at an interesting time, because the U.S. probe Mariner 9 had just reached the neighbourhood of Mars and had been out into a closed orbit round the planet; this was the first time that anything of the sort had been done, since the earlier Mars Mariners (4, 6 and 7) had been fly-by encounters, and the Russians had had no luck at all. When Mariner 9 arrived it could see very little apart from a few 'spots', which we now know to have been the tops of the giant volcanoes poking out from below the level of the obscuration. Only when the dust-storm cleared away was Mariner able to see the true surface, and we had our first views of the Martian landscape.

We have learned to expect dust-storms when Mars is near perihelion and also near opposition. This will happen in 2003; perihelion is due on 30 August and Mars will be at its closest to the Earth on the 27th. I will be very surprised if no dust-storm occurs. Of course, storms must also be whipped up during all perihelion passages, but if Mars is then a long way from opposition it is much less easy for us to follow them – though now, of course, we can use the results from the Hubble Space Telescope, which are much more detailed than any views from the surface of the Earth can possibly be.

Next, let us turn to the polar caps, which are the most striking features on the whole of Mars when they are best placed. The story here goes back to 1666, when observations of Mars were being made by Giovanni Cassini. He saw surface markings well enough to draw them, and also saw that the polar regions were covered with bright white caps, so that they gave every impression of being snow-covered. There was a definite seasonal cycle; the caps were largest in the winter of the hemisphere concerned, and smallest in the summer.

A sketch made by Christiaan Huygens in 1656 is said to have indicated the caps, but to be honest I can make little of it, because all it really shows is a dusky band across the centre of the disk which appears to be an optical effect. However, Huygens certainly saw the south polar cap in 1672, and all later observers have recorded them. In fact, they cannot possibly be overlooked when they are at their best. They are not centred on the exact geographical poles, as Giacomo Maraldi pointed out as long ago as 1719. Moreover, they are not alike. The greater range of temperature in the southern hemisphere means that the south cap can become larger than its counterpart, but also that it can become even more reduced in the height of summer. Recent measurements have shown that the remnant of the south cap, when at its smallest, is centred on a point 250 miles from the pole, while with the north cap the difference is only about 40 miles. In midwinter the south cap may extend down as far as latitude –45 degrees.

Early observers, such as Cassini and Maraldi, said little about the nature of the caps, and were content to do no more than follow their changing aspects. William Herschel was more forthcoming, and in 1784 he made the suggestion that the caps were composed of ice or snow. This, of course, was perfectly logical, if only because the Earth has polar caps of such a kind. For many years nobody seriously questioned the snow-and-ice picture, and in what may be termed the final pre-Space Age period of Martian exploration, Antoniadi stated categorically that 'this theory of Herschel's is correct'.

What was not so clear-cut was the depth of the frozen layer, since it was held that a thick ice-cap would melt and release a great deal of water vapour into the atmosphere – more, indeed, than could be regarded as probable. When a cap shrinks with the arrival of warmer weather it does so quite rapidly, and yet spectroscopes indicated that the Martian atmosphere was virtually bone-dry. A thick cap covering some 4,000,000 square miles (as the southern cap often does) would be expected to make the atmosphere decidedly wet when melting, and this did not seem to fit the facts. As recently as 1954, Gérard de Vaucouleurs wrote that the cap thickness could not be more than a couple of inches, and many astronomers dismissed it as being nothing more than a deposit of hoar-frost, in which case the Martian caps could not be compared with those of our own Antarctica or Greenland. There was also a suspicion that in view of the low atmospheric pressure, the shrinking of a cap could well be due to sublimation (i.e. changing directly from solid into gas) rather than to conventional melting.

Strenuous attempts were made to follow the polar caps through their cycles. It was found that when a cap starts to shrink, its outer rim becomes irregular. With the southern cap, two dark rifts appear, the Rima Australis and the Rima Angusta, and subsequently the cap outline becomes so distorted that a bright white area is left behind, showing up as distinct from the main mass; Schiaparelli, who followed it in 1877, named it the Novissima Thyle. In 1911 the observers of the Mars Section of the British Astronomical Association, at that time directed by Antoniadi, saw it well before it was left behind in the general shrinkage, appearing in the guise of a sparkling white spot well within the main cap; it then showed as a brilliant promontory before the principal cap shrank away and left it isolated. At a later stage in the seasonal cycle, the Novissima Thyle itself usually breaks up into small white dots before vanishing. There are also the so-called 'Mountains of Mitchel', first seen by O. M. Mitchel at the Cincinnati Observatory in 1845, which appear as isolated spots at around latitude −73 degrees when the cap is decreasing, but which do not last for more than a week or so. It was generally supposed that these bright spots, like Novissima Thyle, were seen because the snow or ice coating persisted there for some time after it had melted (or sublimed) from the lower-lying areas nearby.

The changes in the northern cap were thought to be similar, and here too there was the regular appearance of a detached white area, Olympia. With each cap, autumn was characterized by the appearance of a whitish overlying haze which sometimes became bright enough to be mistaken for the cap itself, and which hid the growth of the true cap from our inquiring eyes.

Another phenomenon which caused a great deal of argument was the so-called Lowell Band. In his book *Mars and its Canals*, Lowell discussed the theory that the caps might be due to solid carbon dioxide, and wrote: 'At pressures of anything like one atmosphere or less, carbon dioxide passes at once from the solid to the gaseous state. Water, on the other hand, lingers on in the intermediate stage of a liquid. Now, as the Martian cap melts it is surrounded by a deep blue band which accompanies it in its retreat, shrinking to keep pace with the shrinkage of the cap . . . This badge of blue ribbon about the melting cap, therefore, shows conclusively that carbon dioxide is not what we see, and leaves us with the only alternative that we know of: water.'

Assuming that the band were real, Lowell's argument was theoretically sound, because moist ground does look darker than dry ground, and carbon dioxide deposits could not possibly show a deep blue band. On the other hand, the reality of the band was questioned. Nobody doubts that the band is visible – I have often observed it myself – but is it due to nothing more significant than contrast between the whiteness of the cap and the much less reflective surface beyond the edge? Antoniadi thought so, and in 1930 wrote that the band was illusory, as he had confirmed by 'noting that the band, due to a contrast effect, does not obey the laws of perspective, and that it cannot be photographed'. (Yet Antoniadi also wrote that the brightness of the polar cap might well be reduced near the edges by the presence of bushes and grasses, which sounds quaint now!)

Arguments continued over the years, some of them quite heated. Generally, it was felt that for once Lowell was probably right and Antoniadi wrong. In 1939 de Vaucouleurs made a long series of observations at Le Houga, and decided that even when all possible contrast effects had been taken into account the band was still too dark to be anything but genuine. Later, G. P. Kuiper wrote that 'I observed the band with the 82-inch reflector at McDonald Observatory, Texas, under excellent conditions in April 1950, and found it black . . . The rim is unquestionably real; its width is not constant, and its boundary is irregular.'

Even temporary lakes or marshes at the boundary of the melting polar cap were thought to be possible, though Audouin Dollfus tended to regard the band as merely part of the famous 'wave of darkening' which was then accepted as a seasonal phenomenon. But – were the polar caps made up of ice and snow at all?

Kuiper, writing in 1949, was in no doubt whatsoever. 'The Mars polar caps are not composed of carbon dioxide, and are almost certainly composed of H_2O frost at low temperature, much below 0°C.' The idea of solid carbon dioxide caps had been proposed half a century earlier by two British astronomers, A. C. Ranyard and Johnstone Stoney, but had never become popular, and by the 1950s it had been more or less relegated to the limbo of forgotten things. Neither was there any marked support for the unattractive Kiess theory of solid nitrogen tetroxide. The whole question seemed to be more or less decided.

It is strange how ideas change. The results from the early Mariner fly-by missions, from 1965 to 1969, caused a complete swing of the pendulum. Gone were the water ice or snow caps; instead, it was thought more probable that solid carbon dioxide was the answer. Then, in 1976, came the Viking missions, and the pendulum swung back once more.

It is now clear that at each pole there is a residual cap which is overlaid by a seasonal cap of solid carbon dioxide. The carbon dioxide condenses out of the atmosphere in Martian autumn, producing clouds which accumulate over the poles and effectively prevent us from seeing just how the polar caps develop. When these 'hoods' vanish, the caps below are revealed. With the onset of spring and summer, the temperatures rise and the seasonal caps disappear, leaving only the permanent residual caps below.

The two caps are not alike, because of the differences in climate between the two hemispheres. The northern seasonal cap is smaller and darker than its counterpart in the south, because it is laid down at a time in the Martian year when the atmosphere contains a good deal of dust; this dust is precipitated on to the surface together with the carbon dioxide, whereas the southern cap is laid down at a time when the atmosphere is much less dust-laden. The northern residual cap is the larger of the two, and has a diameter of around 600 miles as against a mere 220 miles for the southern cap. The temperatures differ; the residual southern cap is, predictably, the colder of the two, with temperatures going down to below –110 degrees Centigrade, while temperatures above the residual northern cap have been known to rise to –68 degrees Centigrade, which is well above the frost point of carbon dioxide and not far from the frost point of water in a thin atmosphere which contains only a small amount of precipitable H_2O. Finally, there are very important differences in composition. The northern residual cap is almost certainly water ice, while the southern is a mixture of water ice and carbon dioxide ice.

The fact that carbon dioxide condenses out of the atmosphere during the formation of the seasonal caps leads to a definite reduction in the ground atmospheric pressure. This was first measured by the Viking probes, which made controlled landings on Mars in 1976. I will have more to say about it later.

What about the thickness of the caps? There is no longer any doubt that the residual caps are substantial; we have come a long way from the time when it was widely believed that a cap was nothing more than a wafer-thin layer of hoar-frost. One thing is certain. There can be no liquid water on Mars, because of the low atmospheric pressure, but at least future colonists will have plenty of ice, and in this respect Mars is far more co-operative than the Moon.

There was also the all-important problem of the nature of the surface away from the polar caps. There was little argument about the ochre tracts, which were generally termed deserts (a term due, as noted, to W. H. Pickering in 1886). Not that there were any serious suggestions that a Martian desert might be a kind of Sahara, with oases, palm-trees, and camels wandering about to admire the view – even though Lowell did compare the colour of the Martian tracts with that of the Painted Desert of Arizona, and the similarity is undoubtedly striking. Subsequently, it was suggested that the oxygen originally present in the Martian atmosphere had combined with the surface rocks, producing a layer of iron oxide – in other words, rust. As Rupert Wildt pointed out in 1934, this would explain both the ochre colour and the present scarcity of free oxygen in Mars' atmosphere.

Simple telescopic observations could shed little further light on the problem, but during the oppositions of the 1920s the question was studied by Bernard Lyot, one of the greatest planetary observers of the present century. Lyot's method was to analyze the light from Mars and compare it with moonlight. He also studied mixtures of grey, brownish and bluish volcanic ash, from which he came to the conclusion that both the Moon and Mars were coated with a layer which was, essentially, volcanic ash. Later on, work by Kuiper in America, Dollfus in France, and others led to the general acceptance of coloured minerals such as felsite or limonite.* This, at least, has been confirmed by the space-probes, so that for once we have a cherished theory which has not been thrown overboard.

---

* For the benefit of those interested in chemical matters, felsite is a rock formed of orthoclase (aluminium and potassium silicate) with quartz grains in occlusion, while limonite is a sedimentary deposit of hydrated iron oxide with the formula $Fe_2O_3.3H_2O$.

Where everyone went completely wrong was in the assumption that Mars must have a surface which was no more than gently undulating. I cannot resist yet another quote from de Vaucouleurs, this time dating from 1950: 'If there are any mountains on Mars they can scarcely exceed five or six thousand feet in height, and must be more like ancient plateaux than well-marked chains of massive peaks in sharp relief.' This does not fit in at all well with the towering volcanoes shown by Mariner 9 and the Vikings. Another mistake was in supposing the dark regions to be old, depressed sea-beds, because some of them, notably the Syrtis Major, have been found to be lofty.

So far as the dark areas themselves were concerned, the first challenge to the original ocean theory seems to have come from Schiaparelli as long ago as 1863, when he pointed out that the dark regions did not reflect the Sun's image as sheets of water might have been expected to do. (Much later on, V. Fesenkov, a Russian astronomer who specialized in planetary research, calculated than any water surface with a diameter greater than 300 yards ought to betray itself in this way.) As time went by the oceanic theory was abandoned, and most people accepted Liais' suggestion that the dark areas were tracts of vegetation. Pickering's discovery of detail in the dark regions as well as the ochre deserts seemed to provide confirmation. Other observations also fitted in – notably the famous (or, in my view, notorious) 'wave of darkening', plus undeniable modifications in the outlines of some of the dark areas.

One area which is variable is the patch which Schiaparelli called the Solis Lacus or Lake of the Sun, an elliptical area some 500 miles long and 300 miles wide, with its longer axis lying east–west. It was thus drawn by Maraldi in 1704, and so far as we can tell from the incomplete records it stayed the same until 1926, when the longer axis was found to lie north–south. Later in that year Antoniadi, using the Meudon refractor, drew it as three separate patches, the central one divided from its companions by a dusky 'bridge'. By 1930 all was normal once more, with the longer axis back in its old east–west direction. In 1939 there were fresh changes, and at one time it was seen that the Solis Lacus was made up of a number of small dark spots contained in a generally dusky area. I have been watching the region for the past thirty years, and there seems little doubt that the changes are real. Neither is the Solis Lacus unique in showing variations.

Then, too, there are the reported changes in colour. Quite apart from the 'wave of darkening', it has been claimed that the dark areas are subject to

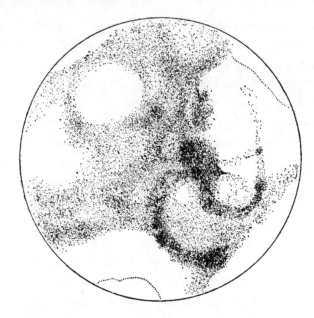

General aspect of the Syrtis Major region, 13 November
1911: Antoniadi (33-inch refractor).

seasonal alterations of hue. Agreement between different observers has never
been complete; thus Antoniadi maintained that the Syrtis Major was bluish-
green in winter and brownish in early summer, while in 1964–65 C. F. Capen,
at the Lowell Observatory, wrote that 'the Syrtis Major was changing from
a blue-green to a green-blue hue . . . The Mare Acidalium changed from its
winter shades of variegated grey and brown to its spring coloration of dark
grey and blue-grey shades with grey-green oases . . . In late spring and early
summer the Acidalium . . . became a very dark grey general shade with a
black-green central area and large grey-green oases'. (Of course, the term
'oasis' is conventional only, and in no way marks a return to Lowell's
Martians.)

Obviously it is hopeless to rely upon one's eyes for precise estimates of
colour when the intensity is so low. I can make no useful contribution, because I admit that I have never been able to see any real coloration in any of
the dark regions, but I do not think that my eyes are particularly sensitive.

An ingenious argument in favour of the vegetation theory was proposed
by E. J. Öpik in 1950. I give it here because it is a classic example of a
theory which appeared too convincing to be anything but true, and which
has nevertheless been shown to be absolutely wrong. Öpik pointed out that

wind-speeds on Mars are appreciable, and that there are major dust-storms. If, therefore, the dark areas were not due to something which could grow and push the dust aside, it should take only a few centuries for them to become completely covered, giving Mars a monotonous, uniform hue. In pre-Mariner days I think that most people regarded this as conclusive evidence that the dark regions were due to organic material of some kind.

Final proof might, it was thought, come from spectroscopic research. If organic matter could be detected in the spectra of the dark patches, the problem would be solved, and therefore efforts were made to detect chlorophyll, the green colouring matter of so many Earth plants. The results were negative. Chlorophyll looks green because the green light is not absorbed, but is reflected back again; it also reflects infra-red wavelengths, which cannot be seen visually, but which may be recorded on infra-red photographs. It was reasoned that if the Martian areas contained chlorophyll, they should appear bright when photographed in infra-red. Disappointingly, they still looked dark.

On the other hand, G. A. Tikhoff in the U.S.S.R. – founder of what has become known, perhaps futuristically, as the science of 'astrobotany' – pointed out in 1960 that not all forms of vegetation could be expected to show evidence of chlorophyll, and that the spectra of the Martian dark areas could reasonably be compared with the spectra of plants living in very cold regions of the Earth, such as the Siberian tundra. After a long series of experiments he decided that while the detection of chlorophyll would admittedly have proved the existence of Martian organisms, the absence of chlorophyll would not necessarily disprove it. For a while, in 1939, it was thought that spectra taken by W. M. Sinton in America really did indicate traces of organic matter – but then, alas, it was found that the observations had been misinterpreted, and the whole question was thrown wide open again. Therefore, it was pointless to speculate about details of the Martian organisms, since it was by no means certain that they existed at all. Lichens were suggested; so were leafed plants of unknown variety, but all that could really be said was that any Martian life must be of very lowly type.

Various theories were proposed in efforts to explain the dark areas without involving life of any kind. One of these was due to Svante Arrhenius, Swedish winner of a Nobel Prize, in 1912. Arrhenius supposed the dark regions to be overlaid with hygroscopic salts, i.e. salts which can absorb moisture, so that they would darken as soon as they picked up the water vapour wafted from the melting polar caps. The idea was more or less forgotten, revived in 1947 by A. Dauvillier in France, and then rejected again, mainly because it seemed that there was not enough moisture in the Martian atmosphere to make the process even remotely possible.

Next came a proposal of an entirely different kind, due to D. B. McLaughlin of the University of Michigan. In 1954 McLaughlin published a paper in which he claimed that the dark areas were nothing more nor less than volcanic ash, ejected from active volcanoes and distributed by the winds. After discussing the air circulation of Mars as compared with that of the Earth, with particular reference to the Martian equivalents of our own trade winds, he went on as follows: 'The sharpish ends of features such as the Syrtis Major point into the wind, and are essentially point sources of some dark material that is carried from these points, fanning out because of variable wind direction. If we restrict ourselves to natural phenomena of which we have experience on Earth, the point sources can have but a single interpretation: they are volcanoes whose ash is carried by the winds and deposited in the pattern we see.'

The irregular changes were attributed to outbursts of exceptional violence, and McLaughlin considered that the eruption which caused the change in form of the Solis Lacus in 1926 was as great as or greater than the disastrous explosion of Krakatoa in 1883. The whole idea was certainly ingenious. On the other hand, there was general scepticism about the prospect of active volcanoes on a world such as Mars.

Moreover, during the early 1960s there began to be doubts as to whether the dark areas were low-lying after all. Radar studies indicated that some of them might be plateaux. In fact, there was no real certainty about anything; Mars hid its secrets well, and only space-probes could solve them.

We now know that the dark areas are simply 'albedo features', where the dusty material has been blown away by winds in the thin Martian atmosphere exposing the darker surface underneath. Their outlines are not nearly so sharp as they appear from Earth, and the features upon them are the same as those in the bright areas. In fact, they are not as important as they look.

Before the Space Age, where were we right and where were we wrong? We were wrong about the atmosphere; it is much thinner than we had expected, and it is made up chiefly of carbon dioxide rather than nitrogen. We were wrong about the dark areas; they are not sea-beds, they are not vegetation-filled and by no means all of them are depressions. Our greatest mistake was in believing Mars to be a living world. But before turning to the results obtained from the Mariners, the Vikings and now the latest missions – Pathfinder and Global Surveyor – it may be entertaining to take a closer look at Mars as we see it from Earth. Therefore, I invite you to join me on a 'telescopic tour' of the Red Planet.

CHAPTER 6

Mars Through the Telescope

Anyone looking at Mars through a telescope for the first time is apt to be disappointed, because generally speaking there is little to be seen apart from a reddish disk, a few dark features, and whiteness over whichever pole happens to be turned toward us. Small instruments will show nothing more, even when Mars is close to opposition.

Today, thanks to the space missions and the Hubble Space Telescope, we know that Mars is a world with an immensely varied landscape. The craters, the channels and other delicate features are quite beyond the range of Earth-based telescopes, but I must pause to say something about two old reports which have been widely discussed, even though they are, in my view, rather suspect.

In 1915 J. R. Mellish, a very experienced planetary observer, turned the 40-inch Yerkes refractor toward Mars. This is the world's largest refracting telescope, and is likely to remain so. Mellish wrote: 'There is something wonderful about Mars. It is not flat, but has many craters and cracks. I saw a lot of the craters and mountains this morning with the 40-inch, and could hardly believe my eyes. This was after sunrise, and Mars was high in a splendid sky. I used a magnification of 750.' Subsequently Mellish learned that Edward Emerson Barnard, an American astronomer renowned for his keen eyesight, had seen craters with the 36-inch refractor at the Lick Observatory, but had not published his results because he was afraid that people would disbelieve him and make 'fun of it'.

This sounds conclusive enough, but Mellish's drawings were burned when his observatory caught fire, and were never published. Neither were Barnard's. The honesty of the two observers is unquestioned, but whether they actually saw craters is another matter. I have looked at Mars through the Yerkes telescope and most of the world's other large refractors, and I can see no sign of craters. This is not significant, because as an observer I am vastly inferior to Barnard and probably to Mellish also, but all in all we must regard the claims as 'non proven'. For the moment let us turn to features which really are accessible to telescopes of moderate aperture.

The map given on page 90 was drawn up by the International Astronomical Union before the Space Age, and is oriented with south at the top, as was then customary; only in the 1960s did it become official to put north at the top – to the everlasting confusion of old-fashioned observers such as myself. There is also the question of nomenclature. Schiaparelli's system, modified and extended by Antoniadi, was accepted up to the Mariner era, but has now become largely obsolete. The names had been drawn mainly from mythology, geography or pure astronomy – such as the Syrtis Major (a gulf in Libya), Mare Tyrrhenum (the Tyrrhenian Sea), Mare Acidalium (really 'the Sea of Venus'), Solis Lacus (the Lake of the Sun), Lunæ Lacus (the Lake of the Moon) and so on. Pavonis Lacus was the 'Peacock Lake', named after the southern constellation of Pavo (the Peacock), while Hellas is the proper name for Greece, and Nix Olympica was 'the Olympic Snow'.

Alas, these names are not appropriate. The Mare Tyrrhenum is not a sea, Solis Lacus is not a lake, and Pavonis Mons is one of the highest of all the Martian volcanoes. So after the space-probe revelations the whole system was overhauled, and new classes of features were introduced:

Catena Crater-chain; a line or chain of craters (e.g. Tithonia Catena).
Cavus Steep-sided hollow (Angusti Cavi).
Chaos Distinctive area of broken terrain (Aromatum Chaos).
Chasma Very large linear chain (Ophir Chasma).
Colles Hills (Deuteronilus Colles).
Dorsum Ridge; irregular elongated elevation (Solis Dorsum).
Fossa Ditch; long, shallow, narrow depression (Claritas Fossæ).
Labyrinthus Valley complex; network of linear depressions (Noctis Labyrinthus is the only really major example).
Mensa Mesa; small plateau or tableland (Nilosyrtis Mensæ).
Mons Mountain or volcano (Olympus Mons).
Montes Mountains (Tharsis Montes).
Patera Saucer-like volcanic structure with fluted or scalloped edges (Alba Patera).
Planitia Smooth, low-lying plain (Hellas Planitia).
Planum Plateau; smooth, high area (Solis Planum).

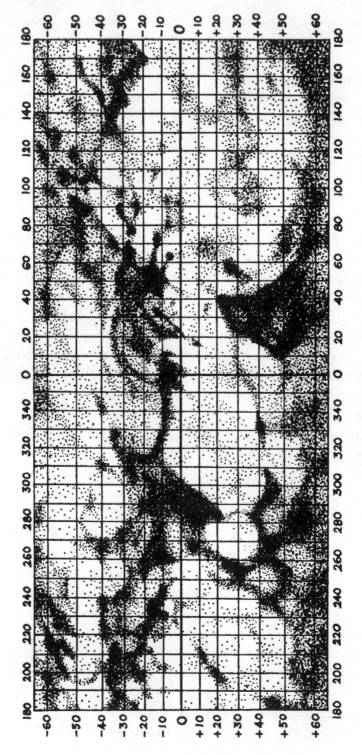

The I.A.U. Map of Mars – the official chart before the Mariner flights.

Acidalium M. (30°, +45°)
Æolis (215°, −5°)
Aeria (310°, +10°)
Aetheria (230°, +40°)
Aethiopis (230°, +10°)
Amazonis (140°, 0°)
Amenthes (250°, +5°)
Aonius S. (105°, −45°)
Arabia (330°, +20°)
Araxes (115°, −25°)
Arcadia (100°, +45°)
Argyre (25°, −45°)
Arnon (335°, +48°)
Auroræ S. (50°, −15°)
Ausonia (250°, −40°)
Australe M. (40°, −60°)
Baltia (50°, +60°)
Boreum M. (90°, +50°)
Boreosyrtis (290°, +55°)
Candor (75°, +3°)
Casius (260°, +40°)
Cebrenia (210°, +50°)
Cecropia (320°, +60°)
Ceraunius (95°, +20°)
Cerberus (205°, +15°)
Chalce (0°, −50°)
Chersonesus (260°, −50°)
Chronium M. (210°, −58°)
Chryse (30°, +10°)
Chrysokeras (110°, −50°)
Cimmerium M. (220°, 20°)
Claritas (110°, −35°)

Copaïs Palus (280°, +55°)
Coprates (65°, −15°)
Cyclopia (230°, −5°)
Cydonia (0°, +40°)
Deltonon S. (305°, −4°)
Deucalionis R. (340°, −15°)
Deuteronilus (0°, +35°)
Diacria (180°, +50°)
Dioscuria (320°, +50°)
Edon (345°, 0°)
Electris (190°, +45°)
Elysium (210°, +25°)
Eridania (220°, −45°)
Erythræum M. (40°, −25°)
Eunostos (220°, +22°)
Euphrates (335°, +20°)
Gehon (0°, +15°)
Hadriacum M. (270°, −40°)
Hellas (290°, −40°)
Hellespontia Depressio (340°, −60°)
Hellespontus (325°, −50°)
Hesperia (240°, −20°)
Hiddekel (345°, +15°)
Hyperboreus L. (60°, +75°)
Iapigia (295°, −20°)
Icaria (130°, −40°)
Isidis R. (275°, +20°)
Ismenius L. (330°, +40°)
Jamuna (40°, +10°)
Juvente Fons (63°, −5°)
Læstrygon (200°, 0°)
Lemuria (200°, +70°)

Libya (270°, 0°)
Lunæ Palus (65°, +15°)
Margaritifer S. (25°, −10°)
Memnonia (150°, −20°)
Meroe (285°, +35°)
Meridianii S. (0°, −5°)
Moab (350°, +20°)
Moeris L. (270°, +8°)
Nectar (72°, −28°)
Neith R. (270°, +35°)
Nepenthes (260°, +20°)
Nereidum Fr. (55°, −45°)
Niliacus L. (30°, +30°)
Nilokeras (55°, +30°)
Nilosyrtis (290°, +42°)
Nix Olympica (130°, +20°)
Noachis (330°, −45°)
Ogygis R. (65°, −45°)
Olympia (200°, +80°)
Ophir (65°, −10°)
Ortygia (0°, +60°)
Oxia Palus (18°, +8°)
Oxus (10°, +20°)
Panchaia (200°, +60°)
Pandoræ Fretum (340°, −25°)
Phaethontis (155°, −50°)
Phison (320°, +20°)
Phlegra (190°, +30°)
Phœnicis L. (110°, −12°)
Phrixi R. (70°, −40°)
Promethei S. (280°, 65°)
Propontis (185°, +45°)

Protei R. (50°, −23°)
Protonilus (325°, +42°)
Pyrrhæ R. (38°, −25°)
Sabæus S. (340°, −8°)
Scandia (150°, +60°)
Serpentis M. (320°, −30°)
Sinai (70°, −20°)
Sirenum M. (155°, −30°)
Sithonius L. (245°, +45°)
Solis L. (90°, −28°)
Styx (200°, +30°)
Syria (100°, −20°)
Syrtis Major (290°, +10°)
Tanaïs (70°, +50°)
Tempe (70°, +40°)
Thaumasia (85°, −35°)
Thoth (255°, +30°)
Thyle I (180°, −70°)
Thyle II (230°, −70°)
Thymiamata (10°, +10°)
Tithonius L. (85°, −5°)
Tractus Albus (80°, +30°)
Trinacria (268°, −25°)
Trivium Charontis (189°, +20°)
Tyrrhenum M. (255°, −20°)
Uchronia (260°, +70°)
Umbra (290°, +50°)
Utopia (250°, +50°)
Vulcani Pelagus (15°, −35°)
Xanthe (50°, +10°)
Yaonis R. (320°, −40°)
Zephyria (195°, 0°)

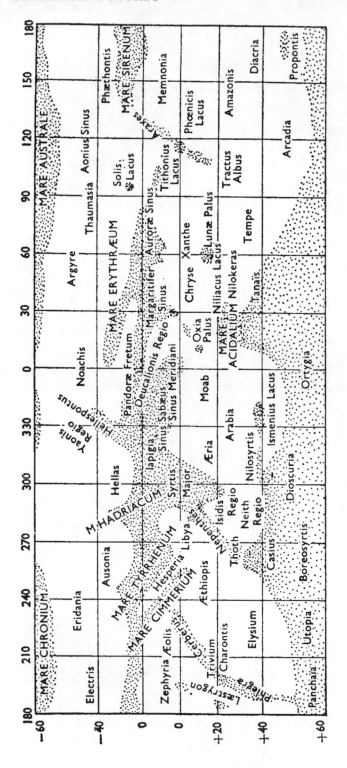

Map of Mars drawn from my own observations in 1963 using my 12½-inch and 8½-inch reflectors.

Rupes Cliff (Ogygis Rupes).
Scopulus Irregular, degraded scarp (Nilokeras Scopulus).
Sulci Intricate network of linear depressions and ridges (Gigas Sulci).
Terræ Lands; names often given to classical albedo features (Sirenum Terra).
Tholus Domed hill (Uranius Tholus).
Vallis Valley; sinuous channel, often with tributaries (Valles Marineris).
Vastitas Extensive plain (Vastitas Borealis, the north-polar circumpolar plain).

My map, given opposite, was compiled from observations made with my own telescopes, of which the largest is now a 15-inch reflector (not used in 1963); I have not added the details which I have seen with large refractors (mainly Lowell's telescope at Flagstaff), so that all the details shown can be detected by anyone with comparable equipment – I do not claim to be at all eagle-eyed. To avoid as much confusion as possible, I have generally used the new names, but older observers at least will be more familiar with Schiaparelli's; for example, Mars Sirenum is more familiar than Sirenum Terra, while Sinus Sabæus (named after the Red Sea) is more likely to strike a chord than Sabæa Terra. Sinus Meridiani, the Meridian Bay, is used as the zero longitude for Mars, and is now officially Meridiani Terra. There is, of course, a connection between the old names and the new, and a minimum of detective work is required to establish which feature is which in either nomenclature. It may be a help to list some of the old and new names:

Old name	New name
Mare Acidalium	Acidalia Planitia
Amazonis	Amazonis Planitia
Aonium Sinus	Aonium Terra
Arabia	Arabia Terra
Arcadia	Arcadia Planitia
Argyre	Argyre Planitia
Ascræus Lacus	Ascræus Mons
Auroræ Sinus	Auroræ Planum
Mare Australe	Australe Planum
Mare Boreum	Boreum Planum
Mare Chronium	Chronium Planum
Chryse	Chryse Planitia
Mare Cimmerium	Cimmeria Terra

Elysium	Elysium Planitia
Mare Hadriacum	Hadriaca Patera
Hellas	Hellas Planitia
Hesperia	Hesperia Planum
Icaria	Icaria Planum
Isidis Regio	Isidis Planitia
Lunæ Lacus	Lunæ Planum
Margaritifer Sinus	Margaritifer Terra
Meridiani Sinus	Meridiani Terra
Nix Olympica	Olympus Mons
Noachis	Noachis Terra
Nodus Gordii	Arsia Mons
Ophir	Ophir Planum
Pavonis Lacus	Pavonis Mons
Promethei Sinus	Promethei Terra
Mare Sirenum	Sirenum Terra
Solis Lacus	Solis Planum
Syria	Syria Planum
Syrtis Major	Syrtis Major Planum
Tempe	Tempe Terra
Tharsis	Tharsis Planum
Mare Tyrrhenum	Tyrrhenum Terra
Utopia	Utopia Planitia
Xanthe	Xanthe Terra

Obviously, the observer has to depend upon the tilt of Mars with respect to the Earth. When the southern hemisphere is inclined toward us, the Syrtis Major is pre-eminent, and there is a line of dark features extending all round the globe – often called the Great Diaphragm. When the northern hemisphere is presented, pride of place generally goes to Acidalia Planitia, though the Syrtis Major is conspicuous even then. As we have noted, the southern hemisphere is tipped toward us at perihelic oppositions, which is why it was better mapped than the northern region until the space-ships took over.

Let us, then, start our telescopic tour. It is convenient to begin with the *Syrtis Major* (I still have an unscientific wish to call it the Hourglass Sea!) which is V-shaped, and extends across the equator into the northern hemisphere. It does show minor changes in outline, but it is always much the

Mars through a very small telescope. I made this sketch in 1939,
using my 3-inch refractor. The polar cap and the Syrtis Major are
quite easy to see. The magnification used was ×150.

most prominent feature on the entire planet, and a small telescope will show
it easily when Mars is well placed. The drawing above was made with my
modest 3-inch refractor.

To the west is the ochre region of *Aeria*, which merges into *Arabia*. To
the east is *Isidis Planitia* (formerly Isidis Regio). Well to the north is *Utopia
Planitia*, celebrated as being the landing site of Viking 2 in September 1976.

Syrtis Major is part of a dark mass which extends for more than half-way
round the planet. It adjoins a rather narrow but often very prominent dark
region, the *Sinus Sabæus*, which is separated from a similar region, *Pandoræ
Fretum*, by the lighter region of *Deucalionis*.

To the south of the Syrtis Major is one of the most famous markings on
Mars: *Hellas Planitia*, which is circular and practically featureless. (In the
'canal' days it was recorded as being crossed by two streaks making up an
X; they were named the Peneus and the Alpheus, but, alas, they do not
exist.) Hellas is very variable in brightness. Sometimes, as in 1967, it can
rival the polar cap; at other oppositions, as in 1995, it is hardly identifiable.
There is no mystery about these changes. When Hellas is bright, its basin is
cloud-filled. Adjoining it, between it and the tract of *Noachis*, is a dark
region, *Hellespontus*; it is heavily cratered, though of course nothing of the

kind was expected before the probe pictures. *Hadriaca Patera*, to the east, is rather similar. (This was formerly Mare Hadriacum – the Adriatic Sea.)

Extending from the Syrtis Major and the adjoining *Libya*, to the east, are two more dark regions, *Tyrrhena Patera* (once the Mare Tyrrhenum) and *Cimmeria Terra* (Mare Cimmerium), separated by the lighter *Hesperia Planum*, which is sometimes – not always – easy to identify. To the north are *Electris* and *Eridania*, two more of the ochre tracts. It was in this area that the capsule of the Russian probe Mars 3 came down in 1971, though without sending back any useful information after its arrival.

Well to the north of the Tyrrhena-Cimmeria streaks is the *Trivium Charontis*, a darkish patch which can be quite prominent at times, and was once thought to be the centre of a system of radiating canals with fascinating names such as Cerberus, Hades, Erebus and Styx. Only the *Cerberus* has any real existence, and it is certainly not a canal.

In this general area are various lightish areas such as *Aethiopis*, *Aetheria*, *Elysium Planitia* and *Isidis Planitia*. On older maps the boundary between Aethiopis and Isidis was marked by one of the most celebrated of the canals, the Nepenthes-Thoth; I have seen a diffuse dusky streak there running in the direction of the wedge-shaped, darkish area of *Casius*.

Now let us go back to the equator. Here we find the Meridian Bay – *Sinus Meridiani*, which has two dark 'forks' pointing northward; it has been called 'Dawes' Forked Bay' and also the Fastigium Aryn, and it marks the zero for Martian longitudes. It extends from the Sinus Sabæus, and under good conditions the forked appearance may be seen with a modest telescope, though it is not always conspicuous and may well be variable in intensity. To its west, and separated from it by a bright region, is the *Margaritifer Sinus*, which is shaped rather like the Syrtis Major, but is much less prominent. Almost due north of it is the *Acidalia Planitia* (Mare Acidalium), with its extension still called the *Niliacus Lacus*, though no doubt the name will be changed before long. Acidalia is the principal dark zone north of the Martian equator, and at aphelic opposition it tends to dominate the scene whenever it lies on the Earth-turned hemisphere.

Between Acidalia and Margaritifer is *Chryse Planitia*, the first site from which signals from Mars were sent back in July 1976. It merges into the similar region of *Xanthe*, and to the west of Xanthe is the dark *Lunæ Planum* (formerly Lunæ Lacus). To the north-west there are ochre tracts such as *Tempe* and *Arcadia Planitia*. Between these two regions the old maps showed

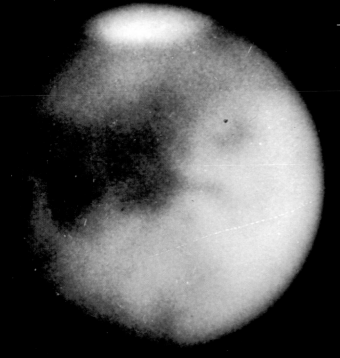

Left: Mars, photographed by C. Capen in 1973 with the 24-inch Lowell refractor.

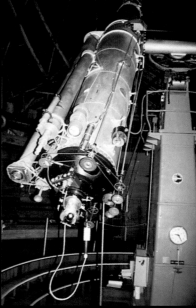

Below: Schiaparelli's map of Mars, showing the canal network; most of the observations were made with the 9-inch refractor at Milan.

Right: The 24-inch Lowell refractor, as I photographed it in 1997. This was the telescope used by Percival Lowell to make his observations of Mars.

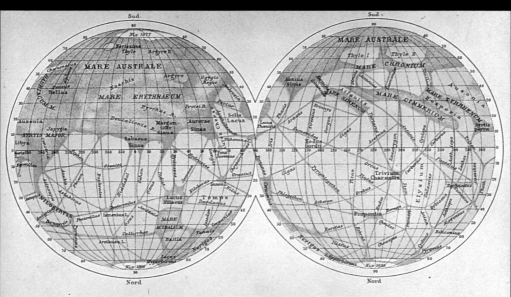

Carte d'ensemble de la planète Mars
avec ses lignes sombres non doublées
observées pendant les six oppositions de 1877-1888
par J.V. Schiaparelli

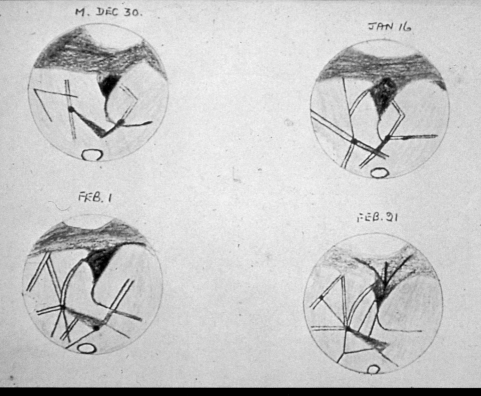

Above: The Martian canals, as drawn by Percival Lowell in 1894.

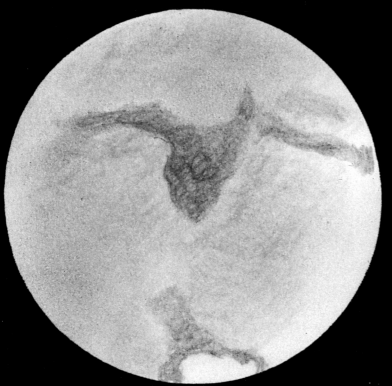

Left: Mars, as I drew it on 4 February 1980 with the Lowell 24-inch refractor, magnification 815. No canals were seen.

Right: Mars 1; the first Russian attempt to send a probe to Mars. Contact with it was lost at an early stage, and was never regained.

Above: Craters, imaged by Mariner 7 in the polar region of the far south. These were nicknamed 'the Giant's Footprint'.

Left: Launch of Mariner 9, the first space-craft to orbit Mars.

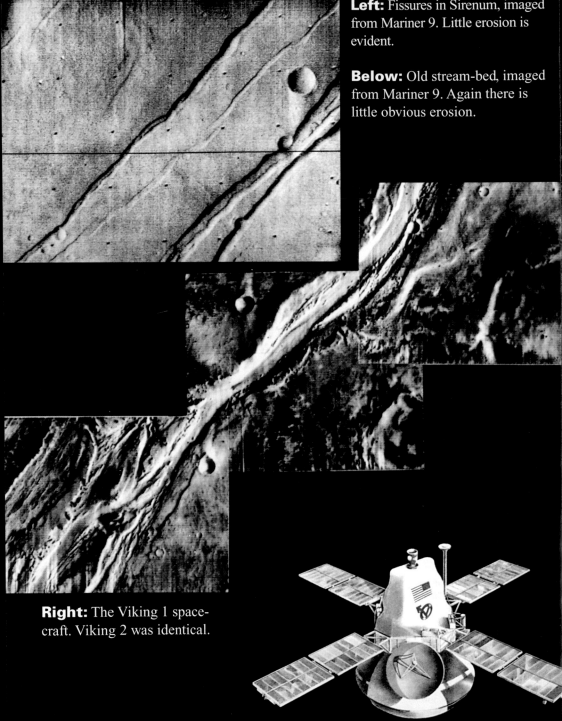

Left: Fissures in Sirenum, imaged from Mariner 9. Little erosion is evident.

Below: Old stream-bed, imaged from Mariner 9. Again there is little obvious erosion.

Right: The Viking 1 space-craft. Viking 2 was identical.

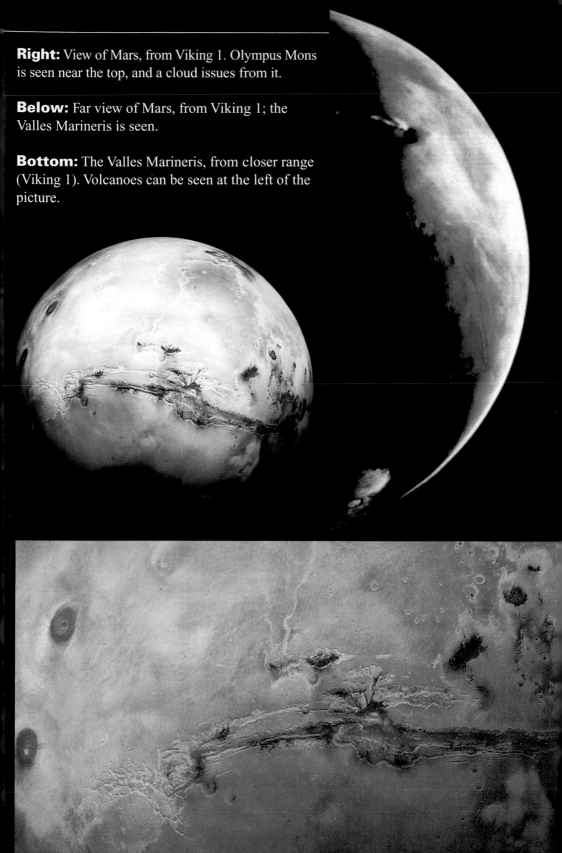

Right: View of Mars, from Viking 1. Olympus Mons is seen near the top, and a cloud issues from it.

Below: Far view of Mars, from Viking 1; the Valles Marineris is seen.

Bottom: The Valles Marineris, from closer range (Viking 1). Volcanoes can be seen at the left of the picture.

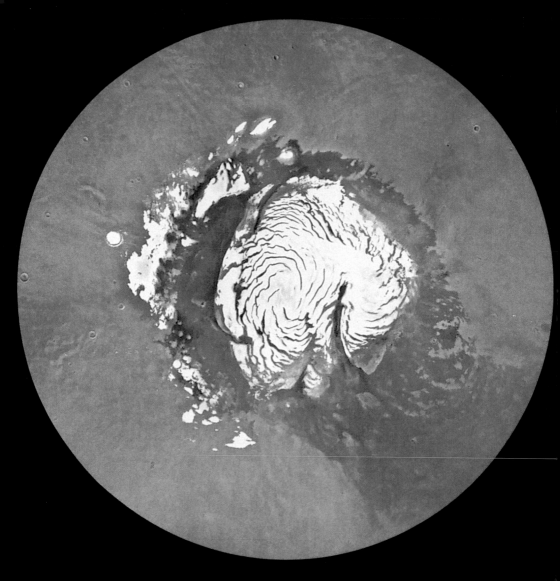

Above: The north polar ice-cap in the Mare Boreum region; Viking image.

Right: Olympus Mons, from Viking orbiter; the loftiest volcano in the Solar System, topped by an immense, complex caldera.

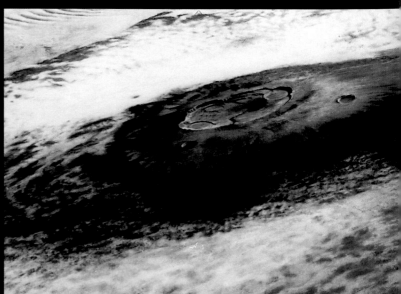

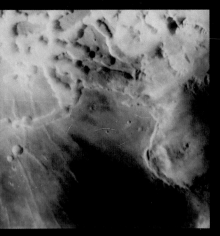

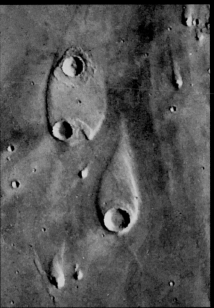

Above left: Morning obscuration in the Noctis Labyrinthus area; Viking orbiter. This is the only really large valley complex of its kind.

Above: Ceraunius Tholus, a typical volcanic structure with a summit caldera.

Above: 'Islands' – almost certainly shaped by the action of running water in the days when Mars had a much denser atmosphere than it has now.

Right: Yuty, a typical Martian crater with a massive central peak. It is 11 miles in diameter, and has high walls. Latitude 22.4 degrees N, longitude 34.1 degrees W.

Above: First colour picture from the Viking 1 lander, in Chryse (the colour correction was later amended).

Right: 'Big Joe', a boulder close to the landing point of VL1. It is very fortunate that the lander did not come down on top of the rock.

a canal, the Ceraunius; this has gone the way of all canals, but at least it has given its name to a lofty hill, the Ceraunius Tholus, which lies in the volcanic area of *Tharsis*.

It is here that we find the greatest of all the volcanoes, the *Olympus Mons*. Also in Tharsis are *Ascræus Mons*, *Pavonis Mons* and *Arsia Mons* – né Ascræus Lacus, Pavonis Lacus and Nodus Gordii respectively. All were recorded from Earth long before probes had become even remotely practicable, and at times they have been seen even when the rest of Mars has been veiled in dust, for the excellent reason that they are high enough to poke out above the dusty layers.

The darkish patch between Margaritifer to the one side and Tharsis on the other is *Auroræ Planum*, which extends toward the sometimes-prominent dark region still called the *Tithonius Lacus*. Here Lowell showed a canal, the Coprates; Mariner 9 revealed the immense Valles Marineris, which is on a giant scale even by Martian standards. Another dark region, the *Mare Erythræum*, lies south of Auroræ; south again is the basin of *Argyre Planitia*, which is of the same type as Hellas, and can be cloud-filled and bright. *Thaumasia* is ochre, and between it and the dark *Phœnicis* is the celebrated *Solis Planum*, which is certainly worth watching. Sometimes it is an easy object, though at other oppositions I have failed to identify it at all even though by all logical standards I ought to have seen it.

There is another prominent dark area in the south-west, *Sirenum Terra* (Mare Sirenum), which has a beak-like extremity; from here Lowell drew a canal, the Araxes, linking Sirenum with Phœnicis, but it too was a mere illusion. *Phæthontis*, yet another ochre region, lies south of Sirenum, with *Memnonia* and *Amazonis Planitia* to its north. Finally, in the north-west adjoining *Arcadia Planitia*, are two darkish patches, *Castorius* and *Propontis*, which I have usually seen without much difficulty when the northern hemisphere has been tilted toward us at a favourable angle.

The polar zones are naturally hard to examine, and there is no point in saying much about them here, but I ought to mention two bright patches in the far south, *Thyle I* and *Thyle II*, as well as the huge northern *Vastitas Borealis* (once Mare Boreum), which extends all round the planet and is covered by the polar cap material during winter.

Again I stress that this is a rough outline guide, in no way a precision chart, but I hope it will help in identifying the main features accessible to the average, well-equipped amateur observer. And, of course, amateurs can make themselves very useful in checking on the clouds, the retreat and form of the polar caps, giving warning of dust-storms and in reporting any unexpected variations in the dark areas. Yet for real knowledge of Mars we must turn to space research, and we begin on 1 November 1962, when the first probe was sent on its way to the Red Planet.

CHAPTER 7

Missions to Mars

Less than a hundred years ago, no man had flown in a heavier-than-air machine. In its way, Orville Wright's brief 'hop' from Kitty Hawk was even more significant than Neil Armstrong's 'one small step' on to the surface of the Moon. Only sixty-seven years separated those two events, and in fact Orville Wright and Neil Armstrong could have met face to face; their lives overlapped.* But certainly the idea of reaching Mars seemed very futuristic in the early part of the twentieth century.

On 4 October 1957 the Space Age opened, not with a whimper but with a very pronounced bang. To the surprise and, it must be admitted, consternation of many people the Soviet Union launched the first artificial satellite, Sputnik 1. It sped round the world, sending back 'bleep! bleep!' signals which to some Western ears sounded faintly derisory. (An American admiral named Rawson earned a place in history by claiming that Sputnik was 'a hunk of old iron that almost anybody could launch', though at that time the United States space programme was floundering helplessly.) Probes to the Moon followed in 1959; Yuri Gagarin made his pioneer ascent in 1961, and it was inevitable that missions to the planets should follow.

In fact, the Russians made an attempt even before Gagarin's flight. There were only two planets within the practicable range: Venus and Mars, and of these Venus had claims to being regarded as the better bet. Only later did we find out that it is overwhelmingly hostile, and in 1961 it was still believed that the surface might be reasonably welcoming, with broad oceans and – possibly – some sort of life. On 12 February the Russians dispatched a Venus probe, but after it had moved out to less than five million miles all contact with it was lost, and although it may have passed within range of Venus in mid-1961 we will never know what happened to it.

---

* Of course I know Neil Armstrong – and I also met Orville Wright when I was learning how to fly at the very beginning of the war. I wonder if, unknowingly, I have also met the first man on Mars?

By this time America's programme was back on course, and two Venus probes were launched from Cape Canaveral during the summer of 1962. The first, Mariner 1, failed; someone forgot to feed a minus sign into a computer, and this makes quite a difference, so that Mariner 1 plunged headlong into the sea. It sank without trace, literally as well as metaphorically. However, Mariner 2 more than compensated for this aberration; in mid-December it made a successful pass of Venus, and sent back the first close-range information from that somewhat sinister world. Meantime, the Russians had turned their attention to Mars.

It is strange that so many people still have completely wrong ideas about space-travel. There is a persistent view that the probe has to 'get out of the Earth's gravity'. In fact there is no possibility of anything of the sort, and no need for it. The Earth's gravitational field weakens with increasing distance, but in theory it has no limit. What has to be done is to work up to escape velocity, so that the gravitational tug is unable to bring the probe back. The only way to achieve this is to use rockets, as was realized in the 1890s by a shy, deaf Russian schoolteacher named Konstantin Tsiolkovskii. In many ways he was decades ahead of his time, even though his papers, published in an obscure journal, attracted very little general interest.

He realized that the main problem would be that of fuel, and so he suggested using several rockets mounted one on top of the other. At launch, the large lowermost rocket would provide the power; when it had used up all its fuel it would break away and fall back to the ground, leaving Rocket No. 2 to carry on with its own motors. When it, too, had run out of fuel the uppermost stage would be put into a path which would take it to its target. In essence this is how all modern interplanetary space-craft function, though needless to say the whole procedure is enormously complicated.

I do not propose to say much about the rockets themselves, because it would be too much of a digression, but we must always remember that a rocket works by what Isaac Newton called the principle of reaction – every action has an equal and opposite reaction. A very easy way to show what is meant is to blow up a balloon and then release it. The compressed air inside rushes out, and 'kicks' the balloon away, making it spiral drunkenly across the room. In a rocket, gases are expelled from the exhaust, and the rocket flies. This works best in vacuum; the rocket is kicking against itself, so to speak, and atmosphere is actually a nuisance, because it sets up friction and has to be pushed out of the way. A space-craft takes off slowly, and accelerates to full velocity only when it is safely beyond the denser part of the Earth's air, and is no longer in danger of being burned away in the same manner as a meteor. Note also that exactly the same principle applies to a Guy Fawkes firework rocket; the main difference is that the gunpowder is replaced by a very complex rocket motor using liquid propellants.

With unlimited fuel supply, there would be no actual need to work up to escape velocity at all, but practical considerations alter the whole situation. What we cannot do is to wait until a planet (such as Mars) is at its closest to us and then simply fire a rocket across the gap. Quite apart from the numerous other objections, this would mean using power throughout the journey, and no vehicle we can build at the moment could possibly carry enough fuel. We must make use of the Sun's gravitational force, and 'coast' for most of the way.

The Earth moves along at a mean velocity of 18½ miles per second, or around 66,000 m.p.h. If it moved faster, it would follow a different orbit, and would initially swing outward; if it were slowed down by some miraculous means, it would at first swing inward. For a Venus probe, then, the principle is to take the vehicle up in a rocket launcher and then slow it relative to the Earth, so that it enters a transfer orbit and comes within range of Venus. But let us concentrate on Mars, where the probe has to be speeded up.

The diagram given below shows the path of Mariner 4, because this was the first Mars probe to be successful. It was launched from Cape Canaveral on 28 November 1964. The massive, compound vehicle carrying the Mariner rose majestically into the air with what seemed to be agonizing slowness; the lowermost rocket – an Atlas – fired its motors, soon shedding two

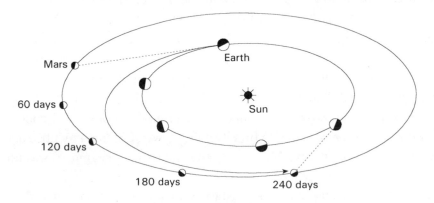

Flight path of Mariner 4.

of its 'boosters' and using the remaining one. When its work was done, Atlas broke away and fell back into the sea. Rocket No. 2, an Agena type, took over at a height of 100 nautical miles and a velocity of 13,000 m.p.h. After another forty-one minutes the Agena fired again, and when it finally shut down for good the space-craft was moving at 25,598 m.p.h. in a path which would take it to within striking distance of Mars. By now it was on its own; Agena, like Atlas, had broken away and fallen back to Earth.

Basically, no more power was needed for the main journey. Mariner was in a transfer orbit, and if the calculations were right the probe would reach the orbit of Mars to rendezvous with the planet. This duly happened. Within two days the Mariner was half a million miles from Earth, and it coasted along until the Mars rendezvous on 14 July 1965 at a mere 6118 miles from the planet. The time elapsing between launch and rendezvous was 228 days, or 222 sols, and the total distance covered was approximately 330,000,000 miles. Recalling that Mars can approach the Earth to within 35,000,000 miles, this sounds a long way – rather like driving from London to Bognor Regis by way of Edinburgh – but in terms of fuel it was the most economical path.

Of course, I have over-simplified matters grossly. A mid-course correction had to be carried out on 5 December 1964; a command was sent to the probe, and was duly obeyed. There was also the question of 'attitude', because if the Mariner pointed in the wrong direction it would be unable to communicate or to receive proper instructions. The method of achieving this was most ingenious. The star Canopus was used as a 'lock', as well as the Sun.

The best way to explain this, I think, is to picture a weight which is hung from a long cord. It will tend to spin, but a second cord, at approximately right angles, will steady it. A line of sight on Canopus was used as Mariner's second 'cord'. Canopus, the second brightest star in the sky, was eminently suitable; it was in the right position relative to the Sun, and its brilliancy enabled the Mariner sensors to locate it, though admittedly only after a day of searching around. The method worked splendidly, and even when the Canopus lock was temporarily lost, at the time of the December mid-course correction, the sensor was able to find it again. Power, obtained by utilizing the solar paddles which gave Mariner its characteristic appearance, was no real problem.

Mariner 4 made a single fly-by of Mars. Though its main task was then over, it did not leave the Solar System; it was moving in a stable path round the Sun, so becoming a tiny artificial planet, and there is no reason to suppose that it will not continue in its path indefinitely, though all contact with it was lost when its power eventually gave out.

I have gone into some detail about the Mariner 4 path because it is absolutely typical, but, as we have noted, Mariner was not the first Mars probe – and this brings us back to 1 November 1962, when the Russians made their pioneer effort.

In those early days of space research the Soviet teams had consistent trouble with their long-range communications, and even at the present time (1998) they have had very little luck with Mars. Their first probe, Mars 1, was quite a massive vehicle, weighing almost 2000 pounds, and apparently it was put into the correct orbit, so that all seemed to be well. The path which would take it to Mars was of the same general type as those of the later Mariners, and it carried a variety of instruments, including several cameras.

Like all other probes, too, Mars 1 was designed to carry out studies of the conditions in interplanetary space. Astronomers were – and still are – interested in the so-called solar wind, which is made up of low-energy atomic particles ejected from the Sun in all directions. Also, the magnetic fields in space are of tremendous theoretical importance; cosmic radiation is another field in which probes are invaluable, and, of course, there is the question of meteoritic particles. Not so many decades ago it was still thought possible that any space-ship daring to leave the protective screen of atmosphere round the Earth would be promptly and fatally battered by a concentrated bombardment of meteoroids. Luckily the danger has been found to be negligible, but there are plenty of micro-meteorites, too small to cause any damage to a space-probe but large enough to be recorded. All the vehicles to the planets have sent back details of the number of hits.

By mid-March 1963 Mars 1 was almost 70,000,000 miles from the Earth, and still going well. Then, abruptly, contact with it was lost, and was never regained. What presumably happened is that the 'star lock' failed, so that the probe swung round and was unable to continue sending or receiving messages. In all probability it passed Mars at around 193,000 miles on 19 June 1963, and there is little doubt that it is still orbiting the Sun, but its fate will never be known. By the time it 'went silent', it was so remote that signals from it, moving at the velocity of light, took twelve minutes to reach the Earth.

Chronologically, Mars 1 was the fourth planetary probe. Earlier vehicles had been Russia's Venera 1, which also went out of contact before getting anywhere near its target, and the two American Venus probes, Mariner 1 (which was a complete failure; it went out of control as soon as it had been launched, and had to be destroyed) and Mariner 2, which made a triumphal pass of Venus in December 1962. Following Mars 1, the Russians attempted another Venus shot with Zond 1, and again they failed. In November 1964, the centre of activity, so far as Mars was concerned, had swung back to the United States.

Two Mariners had been prepared: Nos. 3 and 4. They were identical, and each was designed to by-pass Mars and send back data, including pictures. The reason for building two probes instead of only one was mainly as a safeguard in the event of failure. As events proved, this was a wise precaution.

As with all space-craft, the time of launching was important. It is essential to use as little fuel as possible, and the velocity needed to reach Mars is least when the Earth launch and Mars rendezvous occur on opposite sides of the Sun. The 'window', or period of time when a launch is practicable, is limited to a few weeks every two years. Absolutely ideal launch conditions would be when the take-off point, the Sun, and the arrival point are lined up, but this hardly ever happens, because the orbit of Mars is appreciably tilted with respect to that of the Earth. The inclination is only 1.9 degrees, but this is quite enough to make a considerable difference.

The 'window' for 1964 fell in November. At noon on 5 November, Mariner 3 was sent up from Cape Canaveral. Americans do not celebrate Guy Fawkes' Day, but it was certainly an unlucky time for Mariner 3, because although the launch seemed at first to be successful the planners soon realized that the flight was doomed. During the first rush through the Earth's dense lower air, the delicate space-craft itself is protected by a shield, which is jettisoned as soon as the main atmosphere is left behind. With Mariner 3, the shield stuck obstinately in position, and the dead-weight meant that the velocity was reduced, so that there could be no hope of the probe reaching Mars. Five and a half minutes after the space-craft had been separated from the Agena stage of the launcher, it was ordered to extend its solar panels, which would enable it to use the Sun's energy to provide power for its various items of equipment. Alas, the solar panels failed too, and without them there was no power. Frantically the planners suspended all scientific operations in the Mariner, kept on trying to persuade the panels to extend, and then decided to fire the motor of the space-craft itself in an attempt to jerk the awkward shield free. Before they could do so, the battery ran down. To all intents and purposes Mariner 3 was dead; it had remained in contact for only 8 hours 34 minutes. Silent and untrackable, it entered a path round the Sun, and no doubt it is orbiting even now.

The space-planners were disappointed, but not dismayed. They managed to identify the cause of the problem, and the back-up probe was modified accordingly. By 28 November, Mariner 4 stood on its launching pad; the usual procedure was followed, and this time there was no mishap. The shield dropped away, the solar panels worked perfectly, and the first American messenger to Mars was well and truly on its way. On the following 14 July the rendezvous manœuvre was begun, and shortly after midnight G.M.T. on 15 July Mariner made its closest approach to Mars. For the record, the exact time was 01 hours 0 minutes and 57 seconds; the distance from the Martian surface was 6118 miles. By then the picture sequence had been completed, and the first views were received later in the day.

Altogether Mariner 4 sent back twenty-one pictures of Mars, some of which were almost blank while others showed considerable detail. Although only one per cent of the total surface of Mars came under scrutiny, the results provided astronomers with plenty of food for thought, and one of the pictures – the eleventh – was very clear; it showed an area in Atlantis, the ochre region between Cimmeria and Sirenum, and revealed a 75-mile crater. By then, of course, the presence of craters had been established, and more than seventy were finally recorded. This is no place to go into details of how the television techniques were applied, or of the 'computer enhancement' in which the raw pictures were electronically dismembered, cleaned up, and reassembled to bring out the details. Suffice to say that the techniques evolved were more or less new, and the power actually received from the space-craft was a tiny fraction of one watt: to be precise, 0.000000000000000001 watt (if you care to count the number of zeros, you will find that there are eighteen of them). Anything of the kind would have seemed hopelessly futuristic even at the start of the Space Age, less than a decade earlier.

There was an obvious temptation to compare the Martian surface with that of the Moon, but even at that early stage it was clear that to take the analogy too far would be unwise. I well remember a comment made by one of the NASA scientists when asked whether Mars was like the Earth or like the Moon. He paused, and then said 'Well – to me, it's like Mars', which was a very fair summing-up. But there were some really vital conclusions to be drawn immediately. First, the dark areas and the ochre tracts did not seem to be very different except in colour; there were craters in both. Some were seen in the famous dark region of Sirenum, and there were others in Phlegra, Phæthontis, Atlantis and elsewhere. Neither were the boundaries of the dark regions as well-marked as had been expected, and it is probably true to say that these first Mariner 4 pictures finally killed off the vegetation theory which had been regarded as so nearly proved. Also, Mars was not a world with a flattish landscape, and one elevation on the famous eleventh frame was estimated to rise to 13,000 feet (though the existence of giant volcanoes was not then suspected). Yet one inference drawn at the time has been found to be wrong. It was supposed that the formations were very ancient and considerably eroded, which now seems to be the very reverse of the truth.

Despite the excellence of the television pictures – by 1965 standards, that is to say – the most important results concerned the atmosphere, which turned out to be very thin indeed. The method adopted was most ingenious, and has always been called the Occultation Experiment. Shortly after 02.19 hours on 15 July, more than two hours after closest approach, Mariner 4 went directly behind Mars. Needless to say, it could not be seen; to glimpse a space-craft from Earth at such a range would mean using a telescope much more powerful than anything we can build. But radio signals were clear, and just before the actual occultation these signals were naturally coming to us after having passed through the Martian atmosphere. The way in which these signals were affected gave reliable clues as to the atmospheric composition and density – and there was another opportunity as Mariner emerged from behind Mars at 03.13 hours. The occultation itself had therefore lasted for 1¼ hours, while the signal distortion due to the Martian atmosphere made itself evident for about two minutes before occultation and another two minutes after emersion.

The occultation took place on the sunlit side of Mars, between the regions named on the maps as Electris and Chronium – latitude 55 degrees south, longitude 177 degrees east. The signal was picked up again as Mariner came out on the night side above Acidalia Planitia (Mare Acidalium), at 60 degrees north 44 degrees west, just before sunrise over that part of Mars. The fly-by had been brief indeed, and there could be no second chance.

To be candid, astronomers were rather depressed by the results. It seemed that the ground pressure on Mars could be no more than 4 to 7 millibars, or about 1 per cent of the pressure of the Earth's air at sea-level. The main constituent was the heavy gas carbon dioxide, though it was thought possible that there might be a considerable amount of argon as well. This led at once to the revival of the Ranyard-Stoney theory of polar caps made up of solid carbon dioxide rather than ordinary ice, and it was not until the Viking missions of 1976 that the pendulum swung back again.

Mariner 4 had done all, and more, than its makers could have hoped of it. Now, of course, all track of it has been lost, but no doubt it is still in orbit round the Sun, and we can be reasonably confident about its path, particularly as it was re-contacted for a while in 1966. Mariner 4 is completing one circuit of the Sun every 587 days, with a distance ranging between 102,531,000 miles at perihelion out to 146,029,000 miles at aphelion, not very different from the probable orbit of the dead Mariner 3. Undoubtedly it has an honoured place in astronomical history.

I need do no more than mention Russia's Zond 2, another intended fly-by, which was launched two days after Mariner 4 but which was yet another failure. Probably it passed within a thousand miles of Mars around 6 August 1965, but long before then it had ceased to transmit, and it was some years before the Russians made another attempt. Meantime, events had been taking place much nearer home. The Apollo programme had been put under way, and, as many people will remember, the first astronauts landed on the Moon in July 1969. Only a few days after Neil Armstrong's never-to-be-forgotten 'small step' on to the Mare Tranquillitatis, there was more news from Mars, this time from Mariners 6 and 7. (In case you are wondering about Mariner 5, I should explain that it was a successful Venus probe. It by-passed the planet in October 1967, but need not concern us here.)

Mariner 7 was launched more than five weeks after its predecessor, but took only 130 days to reach Mars, as against 156 days for Mariner 6. The distances covered were, respectively, 197,000,000 miles and 241,000,000 miles. The orbits were of the now-conventional transfer pattern; each probe needed only a single mid-course correction, and there was only one serious alarm. This was near the end of the voyages, on 30 July 1969. Mariner 6 was within seven hours of its pass, with Mariner 7 not far behind, and everything seemed to be under control when, with no prior warning, Mariner 7 'went silent'. The Deep Space Tracking Station at one of the key stations,

Hartebeespoort in South Africa, reported loss of signal, and it was believed that the probe had been hit by a meteorite large enough to damage it. Moreover, Mariner 7 had lost its Canopus lock, which explained why it had gone out of contact. Fortunately it was still able to receive commands from Earth, and eventually Canopus was re-located by its sensors, so that signals were picked up once more. It was found that there was a certain amount of damage, but nothing serious, and the trajectory had been changed very slightly – enough to make the closest approach to Mars occur ten seconds later than had been predicted. After a journey lasting for over four months, this was not very much! In the end, the television pictures from Mariner 7 proved to be the more detailed of the two.

Each probe took what are called far-encounter pictures while still a long way from Mars (well over half a million miles in some cases), and these showed the south polar cap excellently, though without any conspicuous dark band round its edge. Also shown was a bright ring – Olympus Mons (Nix Olympica) which had been seen by Schiaparelli and all later observers of the planet, though neither of the 1969 probes could indicate that it is really a 15-mile-high volcano. It was, I suppose, these long-range views which gave the final coup de grâce to Lowell's canals in any form.

Altogether, Mariner 6 took 50 far-encounter and 25 close-approach pictures; Mariner 7, 93 and 33 respectively. The areas covered were not the same, because the first probe passed more or less over the Martian equator while Mariner 7 concentrated upon the extreme south. Among the regions shown in the pictures were Sirenum, Sabæus, Meridiani, Deucalionis Regio and Hellas. Picture resolution was much better than with Mariner 4, partly because of improved techniques and partly because the 1969 space-craft went closer to Mars, passing a mere 2200 miles from the surface.

There is no point in giving further details about the orbits, and it is enough to say that after rendezvous each probe continued in a closed path round the Sun, so adding to the growing swarm of tiny artificial planets. Generally speaking, the results of the missions were much as had been expected. The low atmospheric density was confirmed; Mariner 6 reported that the pressure in the Meridiani region was 6.5 millibars, while Mariner 7 gave a mere 3.5 millibars over the area which had been known as Hellespontica Depressio, indicating that it was not a depression at all. At noon on the equator the temperature was found to rise to +60 degrees Fahrenheit, falling to below –100 degrees at night. Mariner 7 made temperature measurements of the south cap area, and found a minimum of –190 degrees Fahrenheit, which was very close to the frost point of carbon dioxide there. This agreement was taken as strong circumstantial evidence that the cap really was made up of solid carbon dioxide rather than water ice.

One of the most interesting features to come under scrutiny was Hellas, no longer regarded as a plateau, but as a basin. Virtually no details could be seen inside it, whereas the neighbouring regions – including the Hellespontica 'Depressio' – were thickly cratered. Craters were also seen in the regions covered with white polar deposit, and in particular there was a striking picture of a double crater arrangement which was at once nicknamed the Giant's Footprint.

After the Mariners had completed their passes and gone on their way, a sense of anti-climax prevailed. So far as could be made out from the 20 per cent of Mars photographed during the close-range encounters, there were three types of terrain: cratered regions, what were known as 'chaotic' areas containing few craters but many jumbled ridges, and the much less common smooth parts of the surface of which Hellas was the best example. Astronomers tended to dismiss Mars as a rather dull, inert world with no features of striking interest, and it was even said that the surface was more like that of the Moon than had been expected.

These two Mariners more or less killed off the idea of the Violet Layer; there was no trace of it. A search was made for a magnetic field, but with negative results. In fact we do now know that Mars does have a magnetic field, but it is very weak, and the Mariners were unable to detect it.

There was another factor, too. Because of the prevailing view that the caps were made up almost entirely of solid carbon dioxide, and the certainty that much of the atmosphere was composed of carbon dioxide gas, the possibility of any kind of Martian life receded into the background, particularly since there was no longer the slightest chance that the dark regions were vegetation-tracts. I have always thought it rather fortunate that plans for two more Mars probes, Mariners 8 and 9, were already so well advanced that there was no thought of cancelling them.

We now know that by pure ill-chance, the 1969 space-ships passed over some of the least spectacular areas on the whole of Mars. They missed the volcanoes, the riverbeds and the gaping chasms, so that they gave an entirely misleading impression of what Mars is really like. This was not the faults of the planners, but it does show the folly of jumping to conclusions. Within less than three years, all our ideas were destined to change.

CHAPTER 8

Mariner 9 – and Others

The story of our direct exploration of Mars can be divided into several well-defined parts. First came the fly-by probes, beginning with the unsuccessful Mars 1 of 1962 and ending with the admirable Mariner 7 of 1969. Next followed the 'landers', culminating with the triumphs of the Vikings in 1976. There was then a long lull, and several failures before the most recent phase with Pathfinder and Global Surveyor. I have given a full list of all the probes – successful and otherwise – in Appendix IV. We have already dealt with the fly-by vehicles, so come now to Phase Two.

It began with the space-craft of 1971, of which there were four: two American and two Russian. Their aims were not identical. The United States vehicles were designed to go into closed orbits round Mars and map the surface as thoroughly as they could, naturally obtaining a great deal of miscellaneous information at the same time. The Soviet planners were more ambitious; they hoped to carry out soft landings, and to pick up transmissions direct from the planet.

In the event, only one of the four vehicles was successful. This was Mariner 9, which more than made up for the failures or near-failures of the other three. Before the Viking programme, indeed, Mariner 9 was the source of practically all our reliable information about Mars, and I must discuss it in some detail, so it may be as well to clear the way by disposing of the other space-craft first. This is logical enough, because Mariner 9 was actually the last to be launched even though it led the race to arrive.

Mariner 8, sent up from Cape Canaveral on 8 May 1971, did not arrive at all. The second stage of the Atlas-Centaur launcher failed to ignite; the probe made an undignified descent into the sea some 350 miles north-west of Puerto Rico, and that was that.

Next it was Russia's turn, and the Soviet rocketeers sent up Mars 2 on 19 May, following it with Mars 3 on the 28th. As usual, the launching procedures were carried through without a hitch. Both the probes were put into correct orbits, and both duly rendezvoused with Mars more than six months later. They were, incidentally, the heaviest vehicles so far sent there. Each weighed 10,250 pounds, as against 2150 for Mariner 9.

The Russian vehicles were made up of two parts each: an orbiter and a lander. Basically, the idea was to put the combined craft into orbit round Mars and then, at a prearranged moment, separate the landing section, which would descend through the Martian atmosphere and slow itself down partly by rocket braking and partly by parachute, finally coming to rest gently enough to avoid damage. Meanwhile, the orbiter would continue in its stable path, acting as a relay for the lander as well as carrying out investigations on its own account. I will say much more about this scheme when we come to Viking, so there is no need to dwell upon it now, particularly as neither of the Russian vehicles worked properly.

Mars 2 entered its closed orbit successfully on 27 November 1971, after a total journey of 292,000,000 miles. Just before doing so it ejected a capsule which apparently hit Mars about 300 miles south-west of Hellas, in latitude –44.2 degrees, longitude 213.2 degrees. The capsule carried a Soviet pennant. Whether it carried anything else was not stated, and in any case no scientific information was obtained from the landing section of the probe, so that the experiment would seem to have been fairly useless. All the same, the pennant does represent the first man-made object to land on Mars (just as the Russian Luna 2 was the first object to hit the Moon, in September 1959; how long ago that seems!).

On 2 December, Mars 3 followed its predecessor into orbit, and it too ejected a capsule, which came gently down and landed at latitude –45 degrees, longitude 158 degrees, in a rather light-coloured region between Electris and Phæthontis not far from the northern limit of the south polar cap. Within a minute and a half it began to transmit a picture from the Martian surface. Momentarily all seemed well, but then, after a mere twenty seconds, transmission broke off. It was never re-established, and the fragment of picture received in the U.S.S.R. showed no detail whatsoever. The 12-foot lander had entered the Martian atmosphere at 13,500 m.p.h., and there is no reason to doubt that the actual touch-down was accomplished without mishap, but something unpleasant happened immediately afterwards. Remember, a violent dust-storm was raging at the time, and this may have been the cause of the trouble. Soviet scientists suggested that the parachute, which had been jettisoned at 100 feet above the ground, landed on top of the probe. Alternatively, the lander may have toppled over, or sunk into windblown dust. All these theories were put forward, though in view of the later Viking revelations that Mars is rockstrewn I suggest that the luckless lander came down on top of a large boulder and was fatally holed. No doubt we will find out in the future, when the first expedition goes to the area and locates the remains. (I refuse to accept yet another theory – that a little green Martian wandered over to the probe and switched it off!)

To be fair, the two Russian vehicles were not total failures, because their orbiting sections did send back some information. They carried radio telescopes operating at a wavelength of 3.4 centimetres, and found that the temperature half a metre below the ground never rises above a temperature of –40 degrees Centigrade; they also reported that some of the clouds rose to heights of over six miles, that there were mountains rising to 10,000 feet and depressions sinking to 4000 feet below the mean surface level, that the ochre tracts were likely to be covered with sand rather than limonite or some such mineral, that there was no detectable magnetic field, and that the previous estimations of the atmospheric density were correct. Values of between 5.5 and 6 millibars were given, so that the Martian atmosphere seemed to have about 1/200 the density of our own. The estimations of surface heights were drawn from measurements of the amount of atmospheric carbon dioxide lying above the various regions. Most of the results were of the right order, though of course we now know that some of the peaks are vastly higher than 10,000 feet.

By this time Mariner 9, dispatched from the Cape on 30 May, was already in orbit and had started to send back information, though at first it could do no more than show the tops of dust-clouds. (It had also taken pictures of the two Martian satellites, Phobos and Deimos, but I would prefer to reserve these for a later chapter.) Mariner, of course, had no landing section, so that it was 'all orbiter'. On 14 November its rocket motor was fired for 15 minutes 23 seconds, slowing the space-craft down and putting it into its circum-Martian path. The initial revolution period was 12 hours 34 minutes, slightly more than half a sol, but it was then changed to 11 hours 58 minutes 14 seconds, which brought the probe down to a minimum distance of 850 miles from the surface. The very first pictures showed four spots which later proved to be the tops of volcanoes poking through the dust. We recognize them today as Olympus Mons, Ascræus Mons, Pavonis Mons

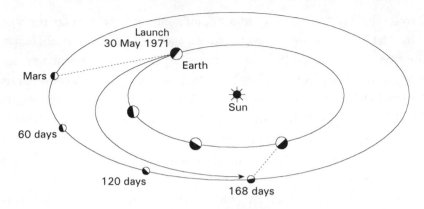

Flight path of Mariner 9.

and Arsia Mons, the giants of the Tharsis ridge. They had been known earlier under the names of Nix Olympica, Ascræus Lacus, Pavonis Lacus and
Nodus Gordii.

When the dust cleared, toward the end of the year, Mariner 9 could begin
its main work. On 30 December the orbit was altered again, giving a new
period of 11 hours 59 minutes 28 seconds. Transmissions continued until 27
October 1972, and altogether Mariner sent back 7329 pictures, covering
practically all of the planet, as against the puny 10 per cent recorded from
the two space-ships of 1969. I doubt whether anyone seriously expected the
pictures to be as good as they actually were; they were breathtaking, and
without them the even more ambitious Viking missions could never have
'got off the ground', either metaphorically or literally.

No doubt Mariner 9 is still circling Mars, though it is now dead. It will
stay there for some time; probably the drag due to the thin Martian atmosphere will eventually make it crash-land, but not for many years. It may well
remain in orbit until after the first colonists arrive.

I have already referred to the 1971 dust-storm on Mars, which was at its
height when Mariner 9 arrived in orbit. It was not unexpected; Mars had
passed through perihelion in September, and had been at opposition in August, so that this ought to have been a very favourable period for Earth-based
observers. In fact the dust hid the surface effectively, and I once noted in my
log that Mars looked like 'an ochre Venus'. We had to wait for the dust to
clear, which took weeks.

Mariner 9, fortunately, could circle patiently and bide its time. It duly
entered its 'mapping path', which carried it from 1019 miles out to 10,440

miles from the Martian surface, and began sending back regular reports. The first features to be shown were, as we have noted, the volcano-tops, Olympus Mons (Schiaparelli's Nix Olympica) and what were at first referred to as North, Middle and South Spots (Ascræus Mons, Pavonis Mons and Arsia Mons). It is worth recalling that during another dust-storm, many years earlier, Schiaparelli had found that his 'Nodus Gordii' and 'Olympic Snow' were almost the only features to be seen. He guessed, correctly, that they must be high, so that we may forgive him for classifying Pavonis and Ascræus as lakes.

Once the Martian atmosphere had cleared, full-scale photographic coverage could begin, and the results were of the greatest interest. The surface proved to be remarkably varied. Since Mariner 9 told us so much about Martian topography, this seems the right moment to go into more detail.

There are high mountains and deep valleys, as well as basins. On Earth we reckon altitudes from sea-level, but there are no seas on Mars – now – and so it has been agreed to use a 'datum line' where the average atmospheric pressure is 6.2 millibars. Olympus Mons, the tallest volcano, rises to almost seventeen miles above this datum line; the deepest canyons go down to well over four miles below it.

The two hemispheres are not alike. Generally speaking, the southern part of the planet is heavily cratered, and much of it lies between half a mile and two miles above the datum line; the northern part is lower – mainly below the datum line – and is much more lightly cratered, so that it must be younger. The very ancient craters which were once there have been eroded away. However, the demarcation line does not follow the Martian equator. Instead, it is a great circle inclined to the equator at an angle of 35 degrees.

The globe of Mars is appreciably flattened; the equatorial diameter is 4218 miles, the polar diameter only 4196 miles. There are also marked local irregularities, and it is rather surprising to find that the most prominent of the depressed basins, Hellas and Argyre, are in the south; the floor of Hellas is over three miles below the datum line, Argyre almost two miles. Both are relatively smooth, and both can become brilliant when cloud-filled – Hellas particularly so, as we have already noted. Gone are the days when Hellas and Argyre were believed to be snow-covered plateaux!

The main volcanic area is the Tharsis bulge. This is a crustal upwarp, centred at latitude 14 degrees south and longitude 101 degrees west; it straddles the demarcation line between the two 'hemispheres', and rises to a general altitude of about six miles. Along it lie the three great volcanoes of Arsia Mons, Pavonis Mons and Ascræus Mons, which are spaced out at intervals of between 400 and 450 miles; only Arsia Mons is south of the equator. Olympus Mons lies 930 miles to the west of the main chain, and is truly impressive, with a 388-mile base and a summit caldera 40 miles across. In the future it will no doubt become a major tourist attraction. I wonder who will be the first man to climb it?

On the northern flank of the bulge is Alba Patera, which is unique on Mars or, so far as we know, anywhere else. It is not lofty, and rises to a height of no more than two miles, but it is over 1500 miles across, with a central caldera.

The second major volcanic area is Elysium, centred at latitude 25 degrees north, longitude 210 degrees west. It is smaller than Tharsis, but it is still over 1200 miles across, and is on average about two and a half miles above the datum line. Volcanoes there include Elysium Mons, Albor Tholus and Hecates Tholus. Isolated volcanoes are also to be found elsewhere – in the Hellas area, for example, as well as near Syrtis Major and Tempe.

The volcanoes themselves are of various types. The 'montes' are large shield structures, essentially similar to our own Hawaiian volcanoes such as Mauna Kea and Mauna Loa; they are much larger and more massive than any comparable volcanoes on Earth, and have gentle slopes of no more than about 7 degrees. The 'tholi' (domes) are smaller and steeper, so that the material from which they formed may have been relatively viscous; like the montes, most of them have summit calderæ. Then there are the 'pateræ', which are scalloped, collapsed shields with shallow slopes and complex summit calderæ; some are symmetrical, but others are less regular, with radial channels running down their flanks. It may well be that they are composed of comparatively loose material, with ash-flows very much in evidence.

The Martian volcanoes are indeed huge. The altitudes of some of them, measured relative to the land adjoining them, are:

Name	Height, feet
Olympus Mons	78,760
Ascræus Mons	59,100
Pavonis Mons	59,100
Arsia Mons	33,600

Elysium Mons	29,500
Tharsis Tholus	19,700
Hecates Tholus	19,700
Albor Tholus	16,500
Uranius Tholus	9800
Ceraunius Tholus	6500

Are these volcanoes active now? Most authorities will say 'no', but we cannot be sure. I will have more to say about this below, when we come to consider the ages of the Martian structures.

Since Mars is a smaller world than the Earth, with a core which is certainly much less hot, we are entitled to ask why the Martian volcanoes dwarf ours. The answer is that plate tectonics are absent there. On Earth, the crust 'slides' over the mantle at a slow but steady rate, and this limits the active life of a volcano. There are what are termed 'hot spots' in the mantle. A volcano forms over one of these, and eruptions continue for a long time, but then the volcano moves away from the hot spot and becomes extinct. We can see this very clearly in Hawaii, where there are two huge volcanoes on the main island: Mauna Kea and Mauna Loa. Mauna Kea has left the hot spot, and all activity has come to an end (at least we hope so, because one of the world's most important observatories has been established at the summit). Mauna Loa has now moved over the hot spot, and is very active indeed, but in time it too will drift away and become dead. Nothing of the sort seems to happen on Mars, so that when a volcano formed over a hot spot it simply stayed there, and was able to build up into a colossal structure.

The greatest canyon system on Mars is that of the Valles Marineris, which is a huge gash in the surface and can be traced for a total length of over 2700 miles, with a maximum width of 370 miles and a greatest depth of over four miles. It begins at the amazingly complex Noctis Labyrinthus, often nicknamed the 'Chandelier', where we find canyons which make our own Grand Canyon seem very puny. It extends eastward toward Auroræ Planum, ending in the blocky terrain not far from one of the most famous of the old albedo features – Margaritifer Sinus (the Gulf of Pearls), now Margaritifer Terra. For part of its course it runs roughly parallel to the line of demarcation, and there are many 'tributaries'.

Craters, of course, are everywhere, many of them with central peaks similar to those of the craters on the Moon. It is now agreed that these are impact structures, formed by cosmical bombardment in the early days of the Solar System. Even the great basins of Hellas and Argyre are attributed to impact, and all in all it is fair to say that Mars has a decidedly battered appearance.

Of even greater significance are the features which look so much like old riverbeds that they can hardly be anything else. There is a saying: 'If it looks like a duck, quacks like a duck and waddles like a duck, then maybe it *is* a duck'. In these old riverbeds we can even see what must once have been islands, and there is evidence of past flash-floods so that some of the old craters have been literally sliced in half. Raging torrents must have carried rocks down into the low-lying areas, and some of the rivers, such as Kasei Vallis, are hundreds of miles long. Several channels debouch into the plain of Chryse, where the landing section of Viking 1 touched down in 1976, and others empty into the Ares Vallis flood-plain south of Acidalia Planitia, between Arabia and Xanthe. This is why Ares Vallis was chosen as the landing site for the Pathfinder probe of 1997. It was thought that there might be rocks of many different types, as did indeed prove to be the case.

Could the channels have been cut by anything except running water? Very liquid lava had been suggested, but can now be ruled out. This means that there was a time when Mars had a much denser atmosphere than that of today, so that there were rivers, ponds, lakes and even seas.

By terrestrial standards, the main Martian features are very old. The past history of Earth is well known, and can be summarised briefly:

Pre-Cambrian era: 3800 to 590 million years ago. Beginning of life; life in the seas.

Palæozoic era: 590 million years ago (beginning of the Cambrian Period) to 248 million years ago (end of the Permian). Sea life; amphibians; reptiles.

Mesozoic era: 248 million years ago (beginning of the Triassic Period) to 65 million years ago (end of the Cretaceous). Reptiles dominant; early mammals. This was the age of the dinosaurs. It is often claimed that these huge creatures died out as the result of a climatic change brought about by the impact of a comet or an asteroid. This may or may not be so, but the theory cannot be ruled out.

Cenozoic era: 66 million years ago (beginning of the Eocene Period) to 2 million years ago (end of the Pliocene). Mammals; primates; ape-men.

Quaternary era: 2 million years ago (start of the Pleistocene Period) to the present day. Ice Ages; true men; civilization.

For Mars the situation is different. Obviously we cannot be as precise as we can be for Earth, but it does seem that there are several reasonably well-defined eras:

Early and Middle Noachian: more than 4.4 thousand million years ago. Primitive ancient crust; early bombardment.
Late Noachian: 4.4 to 3.8 thousand million years ago. Intercrater plains; lava-flows; sinuous channels.
Hesperian: 3.8 to 3.6 thousand million years ago. Lava-flows; complex ridged plains.
Early Amazonian: 3.6 to 2.3 thousand million years ago. Smooth plains such as Acidalia; extensive vulcanism.
Middle Amazonian: 2.3 thousand million years ago to 700 million years ago. Continued vulcanism.
Late Amazonian: 700 million years ago to the present time. Major volcanic activity in Tharsis and elsewhere, dying out at a relatively late stage; disappearance of surface water.

So far as the volcanoes are concerned, it seems that some, such as Elysium Mons and Uranius Tholus, are over 2000 million years old, but some of the volcanoes in Tharsis may have been active well after the beginning of the Late Amazonian, and Olympus Mons last roared a mere 30 million years ago. Mars was then a very different place from the arid, inhospitable world we know.

Yet could these relatively favourable conditions return? And can Mars go through very marked changes of climate in any sort of cycle? It is at least possible, and there are several effects to be borne in mind. And there is one point which has always puzzled me. If the riverbeds, in particular, were very ancient – that is to say, dating back thousands of millions of years – they would surely show effects of long-term erosion; we know that there is plenty of dust in the Martian atmosphere, and that dust is highly abrasive. Surprisingly, many of the old stream-beds look very clear-cut. Of course, I am not suggesting that they are young by conventional standards, and the fact that some of them have craters on their floors shows that water stopped flowing long ago, but I do suggest that they may not be as old as most people believe.

One idea was put forward in 1971 by Carl Sagan, who was firmly convinced that life had once existed on Mars. His theory depended upon the

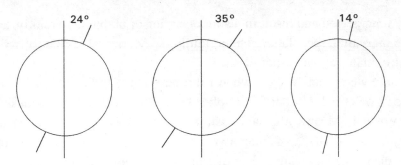

The range of inclination of the axis of Mars. The present value
is 24°, very similar to that of the Earth.

changing tilt of Mars' axis, and there is an analogy here with the Earth. The
Earth is not a perfect sphere; it bulges slightly at the equator, and the pulls of
the Sun, Moon and (to a very minor degree) the planets act upon this bulge,
so that the direction of the axis is not absolutely constant.

In fact the angle of inclination of the Earth's axis does not alter very
much, and of course the Earth's orbit is not far from circular; the fact that we
are slightly closer to the Sun in December than in June makes very little
difference. Not so with Mars, where the orbit is more eccentric, and the
precessional period is of the order of 51,000 years. The Martian axial inclination ranges between 14.9 degrees and 35.5 degrees, so that it is sheer
chance that the present value (23°59') is practically the same as ours. Mars'
orbit is also more elliptical than ours, and the eccentricity ranges between
0.004 to 0.141 in a cycle of 90,000 years; the current value is 0.093.

As we have noted, summer in the southern hemisphere now falls when
Mars is at its closest to the Sun, though in 25,000 years' time the situation
will be reversed, and at perihelion it will be southern winter. But there are
periods when one or other of the poles is tipped much more markedly sun-
ward at a perihelion than is the case now. Could the volatiles then sublime,
thickening the atmosphere and even causing rainfall? Clearly a denser
atmosphere produced in this way would be only temporary, but it might
conceivably make Mars fertile for a brief period.

Another idea is that every few tens of millions of years Mars goes through
spells of intense volcanic activity, when tremendous quantities of gases and
vapours (including water vapour) are sent out from beneath the crust, and
the atmosphere is thickened up for a while. The problem here is in trying to
decide just why this should happen, though activity might build up over a

very long period and result in a sudden series of outbursts. Frankly, all this is no more than speculation, but certainly the Mariner 9 results caused us to modify many of our earlier ideas.

There were many regrets when Mariner 9 came to the end of its career; contact was finally lost on 27 October 1972. It had carried out its task superlatively well. Meanwhile, the Russians had certainly not lost interest, and in the summer of 1973, less than a year after the last signals from Mariner 9, they dispatched a positive fleet of space-ships. Four vehicles were sent up in rapid succession. At least two were of the orbiter-plus-lander type, and much was expected of them, particularly as the Soviet rocket-planners had already had marked success with Venus, which is a far more difficult subject for exploration than Mars.

Sad to say, the Soviet fleet was a failure, and added very little to the information drawn from Mariner 9. To avoid tedious repetition, it may be best to treat the four probes briefly and in order of launch:

Mars 4. Launched 21 July 1973. On 10 February 1974 it approached its target, but its braking engine failed to operate, and the probe missed Mars by over 1300 miles, so that it continued on its way in a solar orbit. A few television pictures were obtained during the involuntary fly-by, but their quality was poor.

Mars 5. Launched 25 July; approached Mars on the following 12 February. This time the braking engine worked, and some television pictures were obtained. They showed craters and riverbeds, but with nothing like the clarity of Mariner 9.

Mars 6. Launched 5 August; reached the neighbourhood of Mars on 12 March 1974. The lander was successfully broken free from the orbiter at a distance of almost 30,000 miles from Mars, and the initial descent seemed promising. Rocket braking was used for five minutes, and then the main parachute was deployed. The parachute phase lasted for 148 seconds, but then contact was permanently lost. So far as is known, the lander came down at latitude –24 degrees, longitude 25 degrees west, between Erythræum and Margaritifer Terra, but nothing more was heard from it. Some television pictures were received before the separation, and the Soviet news agency said that from Mars 6 the planet looked like a red, waning Moon.

Mars 7. Launched 9 August 1973; approached Mars on 9 March 1974, three days before Mars 6. This time the lander separated from the orbiter prematurely, and the vehicle missed Mars by 800 miles, moving uselessly into an orbit which will continue to take it round the Sun.

The story of the Red Fleet is rather depressing, and even now, a quarter of a century later, the Russians have had absolutely no luck so far as Mars is concerned. But before going on with the story, let us pause to summarize what Mariner 9 told us about the topography of the planet.

CHAPTER 9

Mars from Orbit

The maps given here (pages 126–133) are not intended to be anything more than general outlines, and are included mainly for the sake of completeness. Following the official policy I have oriented them with north at the top, but this does not matter from the observer's point of view, because Earth-based telescopes will show none of the details apart from the albedo regions and the main basins, though a few of the volcanoes, notably those in Tharsis, can be glimpsed as tiny specks.

Crater diameters are given, and longest dimensions in the case of linear or extended features. All units are in miles.

North Polar Area, down to around latitude +60 degrees. This is covered with white deposit during winter. There are not very many craters by Martian standards, but some are visible, and one of them (Korolev) was later shown by Viking to be filled with ice. The main feature of the area is the layered or laminated terrain. This is made up of what may be called 'staircases', very thin layers of alternate light and dark hue, whose gently sloping faces show a certain amount of relief, descending from the pole toward the cratered regions. The layers have been likened to saucers of various sizes, and are peculiar to the two polar zones. The huge, dusky Vastitas Borealis extends right round the planet, from approximately latitude +55 degrees to +67 degrees.

North-east Quadrant (see page 126). This quadrant contains the great Syrtis Major, now known to be a plateau sloping off to either side. To the west is one of the few heavily-cratered regions of the northern hemisphere, making up ochre tracts such as Aeria, Arabia and Eden. Sinus Meridiani or the Meridian Bay, to the lower left, was long accepted as the zero for longitude; it is purely an albedo feature, and has a pronged aspect when well seen, so that it is often nicknamed Dawes' Forked Bay in honour of a last-century observer of the planet, the Rev. W. R. Dawes. The currently-accepted zero for longitude is the 35-mile crater Airy-0, named in honour of Sir George Airy, the rather formidable nineteenth-century Astronomer Royal who was largely responsible for the acceptance of Greenwich as the zero for terrestrial longitude. Airy-0 is right on the meridian (because it was so ordained!) but is slightly south of the equator (latitude 5°.2 S).

Isidis Planitia (formerly Isidis Regio), bordering Syrtis Major, slopes down to the volcanic region of Elysium Planitia, where there are several well-marked volcanoes: Elysium Mons, Hecates Tholus and Albor Mons, all of which are decidedly lofty even though they cannot compare with the giants of Tharsis. The relative sparseness of crater distribution around here is very evident. The quadrant also includes Utopia Planitia; it was here that the second Viking Lander came down, not very far from the prominent crater now called Mie. Utopia had been expected to be relatively smooth. Instead it has proved to be what has been termed 'a forest of rocks'.

Some of the most famous of the Lowellian canals – Phison, Euphrates, Protonilus, Hiddekel and others – were in this quadrant. All, alas, were quite illusory and do not correspond to genuine features of any kind.

North-west Quadrant (see page 128). This is the quadrant which contains most of the Tharsis volcanoes; only Arsia Mons is south of the equator. Olympus Mons, of course, is unrivalled; it is a typical shield volcano, crowned by a 40-mile complex caldera. During Martian mornings it is wreathed in clouds formed of water ice, condensed from the atmosphere as it cools while moving up the slopes of the volcano.

Shield volcanoes were originally so called because of a supposed resemblance in shape to the shields of early Viking warriors. On Earth they are confined mainly (not entirely) to three areas: Hawaii, the Galapagos Islands and Iceland, plus a few in California and New Zealand. Summit calderæ are the rule, and this also applies to the Tharsis volcanoes. Ascræus Mons has a 31-mile caldera; Pavonis, 28 miles; and Arsia Mons, which actually lies just in the south-west quadrant, beats them all with a caldera over 85 miles in diameter.

The Tharsis volcanoes are associated with what seem to be drainage systems. The most conspicuous features of the surrounding area are the canyons, mainly the tremendous Valles Marineris and the labyrinth at Noctis, which lie in the south-west quadrant. Another interesting feature is the Tractus Albus, which is ridge-like and which runs from Arsia Mons as far as Acidalia Planitia. It is certainly associated with the diagonal line of the main volcanoes, of which the four giants are not the only members.

The Tractus Albus runs between the Tharsis volcanoes to the east and the darkish Lunæ Patera to the west; it drops in height with increasing distance northward. Acidalia is a low-lying region, and is quite unlike the Syrtis Major even though these are the two most conspicuous dark areas on the whole of Mars.

Tharsis slopes off sharply to the west (Amazonis Planitia), but less sharply toward the east. Here we come to an ochre tract, Chryse Planitia, which merges into the very similar Xanthe. It was in Chryse that the lander of Viking 1 made its epic descent.

South-west Quadrant (see page 130). Two major features dominate this part of Mars: the Valles Marineris, and the basin of Argyre.

The Valles Marineris is huge even by Martian standards, since it is over 2500 miles long, more than 45 miles wide at its broadest point and 20,000 feet deep, which is practically three times the depth of our Grand Canyon of the Colorado. It is in the nature of a rift, and comparisons have been made between it and our Red Sea, though the Valles is considerably the longer of the two. The whole system extends from Tharsis right through to Auroræ Planum, and one has to admit that it does correspond more or less to the Coprates canal shown on the old Lowell-type maps, though this is certainly nothing more than coincidental. To the north-west is another familiar feature, long known as Juventæ Fons or the Fountain of Youth, which is an irregular depression visible from Earth as a dark patch. Lowell, needless to say, regarded it as an oasis.

Adjoining the Tharsis region and the high area of Phœnicis is the Noctis Labyrinthus, formerly called Noctis Lacus. This is a huge system of canyons, unlike anything else on Mars, making up the pattern which has been nicknamed the Chandelier. The canyons are not like the rills of the Moon, and seem to be due to fracturing of the surface, together with the withdrawal of magma from below. Each of the main canyons has an average width of about a dozen miles.

To the east are two famous dark areas, Cimmeria Terra and Sirenum Terra; Cimmeria is rather high, while Sirenum seems to have no special characteristic apart from its colour. There are many craters round here, and the Russian probe Mars 3 came down in a position near one of these, Ptolemæus.

Argyre Planitia (formerly Argyre I, to distinguish it from a smaller basin, Argyre II) is of the Hellas type, though it is neither so large nor so deep. Little can be seen in it; to its north are sinuous features which can be classed as dry riverbeds. To the north-west lie the dark, cratered regions of Erythræum and Margaritifer Terra.

South-east Quadrant (see page 132). Overall, this is one of the most heavily cratered parts of Mars, and there are few true volcanoes, though a couple have been located in Tyrrhena Patera, north-east of Hellas at around latitude −22 degrees, longitude 253 degrees. Magnificent systems of the riverbeds are seen in the Rasena region (latitude −25 degrees, longitude 190 degrees). Among the craters, one of the most extraordinary is Proctor, east of Hellas, which appears to be filled with sand-dunes. It lies on the Hellespontus, a cratered region sloping down toward the Hellas basin itself.

I have already said a good deal about Hellas. It is over 1300 miles across, and apart from one small patch, seen by Antoniadi and named by him Zea Lacus, it is virtually featureless. Across it Schiaparelli, in 1877 and 1879, drew two canals which were later called the Alpheus and the Peneus, making up a cross. No trace of them was shown by Mariner 9, so yet again we seem to be dealing with alleged canals which do not exist in any form whatsoever.

South Polar Area. Here we are back to layered terrain, but the region round the south pole is more thickly cratered than its northern counterpart. There are dark areas at high latitudes − notably Chronium Planum and Australe Planum − but nothing so continuous as the northern Vastitas Borealis.

Comparing these results with those of the earlier probes leaves one in no doubt as to the magnitude of Mariner 9's achievement, which was all the more remarkable because of the need to shoulder the programme originally assigned to Mariner 8. From being regarded as an inert, lunar-type world, Mars was transformed into an active planet, with features which were as fascinating as they were surprising. And as Mariner 9 came to the end of its career, the Vikings were being prepared to take up the torch.

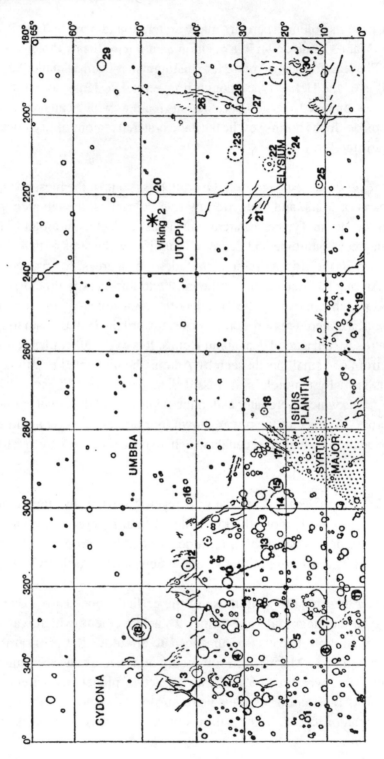

North-east quadrant.

	Name	Lat.°	Long.°W	Diameter	Longest dimension
1	Gill	15.8 N	354.5	50	–
2	Focas	33.9 N	347.2	51	–
3	Deuteronilus Mensa	45.7 N	337.9	–	390
4	Cerulli	32.7 N	337.9	75	–
5	Pasteur	19.6 N	335.5	71	–
6	Henry	11.0 N	336.8	103	–
7	Arago	10.5 N	330.2	96	–
8	Lyot	50.7 N	330.7	136	–
9	Cassini	23.8 N	327.9	258	–
10	Quénisset	34.7 N	319.4	79	–
11	Janssen	2.8 N	322.4	96	–
12	Moreux	42.2 N	315.5	86	–
13	Flammarion	25.7 N	311.7	99	–
14	Antoniadi	21.7 N	299.0	236	–
15	Baldet	23.0 N	294.5	121	–
16	Renaudot	42.5 N	297.4	43	–
17	Nili Fossa	24.0 N	283.0	–	441
18	Peridier	25.8 N	276.0	62	–
19	Escalante	0.3 N	244.8	52	–
20	Mie	48.6 N	220.4	58	–
21	Elysium Fossa	27.5 N	219.5	–	692
22	Elysium Mons	25.0 N	213.0	268	–
23	Hecates Tholus	32.7 N	209.8	114	–
24	Albor Tholus	19.3 N	209.8	103	–
25	Eddie	12.5 N	217.8	56	–
26	Phlegra Montes	40.9 N	197.4	–	813
27	Lockyer	28.2 N	199.4	46	–
28	Adams	31.3 N	197.1	62	–
29	Stokes	56.0 N	189.0	44	–
30	Orcus Patera	14.4 N	181.5	237	–

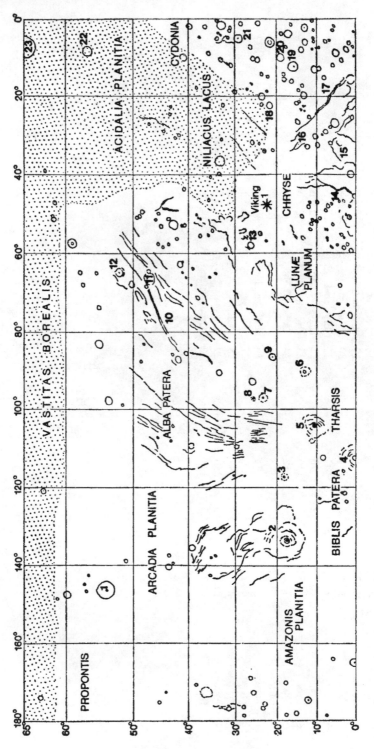

North-west quadrant.

	Name	Lat.°	Long.°W	Diameter	Longest dimension
1	Milankovic	54.8 N	146.6	70	–
2	Olympus Mons	18.4 N	133.1	388	–
3	Jovis Tholus	18.4 N	117.5	38	–
4	Pavonis Mons	0.3 N	112.8	233	–
5	Ascræus Mons	11.3 N	104.5	301	–
6	Tharsis Tholus	13.4 N	90.8	95	–
7	Ceraunius Tholus	24.2 N	97.2	67	–
8	Uranius Tholus	26.5 N	97.8	44	–
9	Fessenkov	21.9 N	86.4	53	–
10	Mareotis Fossa	45.0 N	79.3	–	438
11	Barabashov	47.6 N	68.5	78	–
12	Perepelkin	52.8 N	64.6	70	–

	Name	Lat.°	Long.°W	Diameter	Longest dimension
13	Sharonov	27.3 N	58.5	59	–
14	Shalbatana Vallis	5.6 N	43.1	–	427
15	Simud Vallis	11.5 N	38.5	–	667
16	Tiu Vallis	8.6 N	34.8	–	603
17	Ares Vallis	9.7 N	23.4	–	1050
18	McLaughlin	22.1 N	22.5	56	–
19	Trouvelot	16.3 N	13.0	104	–
20	Becquerel	22.3 N	7.9	104	–
21	Curie	29.2 N	4.9	61	–
22	Kunowsky	57.0 N	9.0	37	–
23	Lomonosov	64.8 N	8.8	94	–

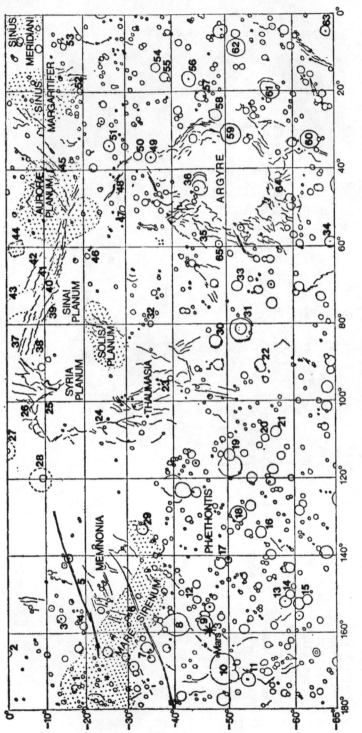

South-west quadrant.

	Name	Lat.°	Long.°W	Diameter	Longest dimension
1	Eiriksson	19.6 S	173.3	35	–
2	Nicholson	0.0	164.5	71	–
3	Burton	14.5 S	156.3	85	–
4	Comas Sola	20.1 S	158.4	82	–
5	Memnonia Fossa	21.9 S	154.4	–	851
6	Sirenum Fossa	34.5 S	158.2	–	1686
7	Mariner	35.2 S	164.3	94	–
8	Newton	40.8 S	157.9	178	–
9	Ptolemaeus	46.4 S	157.5	114	–
10	Copernicus	50.0 S	168.6	181	–
11	Liu Hsin	53.7 S	171.4	75	–
12	Hipparchus	45.0 S	151.1	65	–
13	Wright	58.6 S	150.8	66	–
14	Keeler	60.7 S	151.2	57	–
15	Trümpler	61.7 S	150.6	48	–
16	Clark	55.7 S	133.2	58	–
17	Nansen	50.5 S	140.3	51	–
18	Hussey	53.8 S	126.5	62	–
19	Porter	50.8 S	113.8	70	–
20	Brashear	54.1 S	119.2	78	–
21	Ross	57.6 S	107.6	55	–
22	Coblentz	55.3 S	90.2	69	–
23	Thaumasia Fossa	45.9 S	97.3	–	695
24	Claritas Fossa	34.8 S	99.1	–	1264
25	Noctis Labyrinthus	7.2 S	101.3	607	–
26	Tharsis Montes	2.8 N	113.3	–	1308
27	Pavonis Mons	0.3 N	113.3	233	–
28	Arsia Mons	9.4 S	120.5	301	–
29	Pickering	34.4 S	132.8	70	–
30	Slipher	47.7 S	84.5	80	–
31	Lowell	52.3 S	81.3	125	–
32	Lampland	36.0 S	79.5	44	–
33	Douglass	51.7 S	70.4	60	–

	Name	Lat.°	Long.°W	Diameter	Longest dimension
34	von Kármán	64.3 S	58.4	62	–
35	Nereidum Montes	41.0 S	43.5	–	1042
36	Hooke	45.0 S	44.4	90	–
37	Ius Chasma	7.2 S	84.6	–	623
38	Tithonius Chasma	4.6 S	85.6	–	562
39	Melas Chasma	10.5 S	72.9	–	327
40	Valles Marineris	11.6 S	70.7	–	2566
41	Coprates Chasma	13.6 S	60.7	–	598
42	Ophir Chasma	4.0 S	72.5	–	156
43	Gangis Chasma	8.4 S	48.1	–	336
44	Juventae Chasma	1.9 S	61.8	–	308
45	Eos Chasma	12.6 S	45.1	–	599
46	Lassell	21.0 S	62.4	53	–
47	Ritchey	28.9 S	50.9	51	–
48	Nirgal Vallis	28.3 S	41.9	–	318
49	Hale	36.1 S	36.3	81	–
50	Bond	33.3 S	35.7	65	–
51	Holden	26.5 S	33.0	88	–
52	Spencer Jones	19.1 S	19.8	53	–
53	Beer	14.6 S	8.2	50	–
54	Vogel	37.0 S	13.2	77	–
55	Hartwig	38.7 S	15.6	65	–
56	Lohse	43.7 S	16.4	97	–
57	Helmholtz	45.6 S	21.1	67	–
58	Wirtz	48.7 S	25.8	80	–
59	Galle	50.8 S	30.7	143	–
60	Maraldi	62.2 S	32.1	74	–
61	Darwin	57.2 S	19.2	103	–
62	Green	52.4 S	8.3	114	–
63	Wegener	64.3 S	4.0	44	–
64	Charitum Montes	58.3 S	44.2	–	878
65	Halley	48.6 S	59.2	50	–

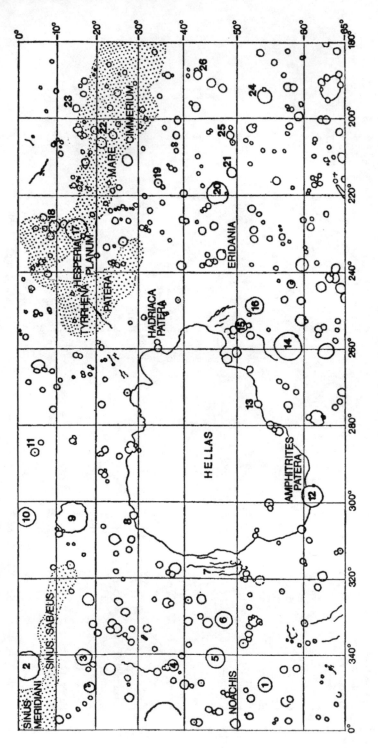

South-east quadrant.

	Name	Lat.°	Long.°W	Diameter	Longest dimension
1	Russell	55.0 S	347.4	86	–
2	Schiaparelli	2.5 S	343.4	287	–
3	Flaugergues	17.0 S	340.0	146	–
4	Le Verrier	38.2 S	342.9	86	–
5	Kaiser	46.6 S	340.9	125	–
6	Proctor	47.9 S	330.4	104	–
7	Hellesponti Montes	45.5 S	317.0		423
8	Niesten	28.2 S	302.1	71	–
9	Huygens	14.0 S	304.4	284	–
10	Schröter	1.8 S	303.6	209	–
11	Fournier	4.2 S	287.5	70	–
12	Barnard	61.3 S	298.4	80	–
13	Gledhill	53.5 S	272.9	45	–

	Name	Lat.°	Long.°W	Diameter	Longest dimension
14	Secchi	57.9 S	257.7	135	–
15	Tikhov	51.2 S	254.1	67	–
16	Wallace	52.8 S	249.2	99	–
17	Herschel	14.9 S	230.1	189	–
18	Knobel	6.6 S	226.9	79	–
19	Martz	35.2 S	215.8	57	–
20	Kepler	47.2 S	218.7	136	–
21	Tycho Brahe	49.5 S	213.8	67	–
22	Graff	21.4 S	206.0	98	–
23	Boeddicker	14.8 S	197.6	67	–
24	Campbell	54.0 S	195.0	76	–
25	Huggins	49.3 S	204.3	51	–
26	Bjerknes	43.4 S	188.7	55	–

CHAPTER 10

The Vikings

Outstanding success though it was, Mariner 9 was no more than a stepping-stone in our exploration of Mars. What we needed next was information sent back direct from the planet's surface. The Russians had tried, and failed; now it was the turn of the United States team. The aim was to soft-land a capsule and hope that it would go on transmitting for many weeks or even months.

Moreover what Mariner 9 had not done, and could not possibly do, was to tell us whether or not Mars was completely sterile. This was one of the main tasks of the Vikings, but let me stress at the outset that it was by no means the sole reason for attempting a controlled landing. Indeed, many astronomers maintained that some of the other investigations, such as atmospheric composition and analysis of the soil materials, were even more important. Opinions differed, but everyone agreed that in any case the Vikings would – if they were successful – open up whole new lines of investigation.

Two missions were planned. Vikings 1 and 2 were identical, and were intended to carry out identical tasks from two different regions of the planet. No. 1 was aimed at the 'Golden Plain', Chryse, more or less between the Margaritifer Terra to the south and the dark mass of Acidalia to the north. No. 2 had as its target Cydonia, an area considerably closer to the north pole, and just about at the limit of the northern polar cap at its maximum spread. The sites had been chosen very carefully. They were relatively low-lying, and were expected to be reasonably smooth; also, they were well placed to receive any residual moisture, because they lay at the end of the great drainage-system associated with the Tharsis volcanoes. If there were any moisture on Mars, it would be expected to be there.

Each vehicle consisted of two main parts, an Orbiter and a Lander. Naturally, the two would travel together across interplanetary space, and would

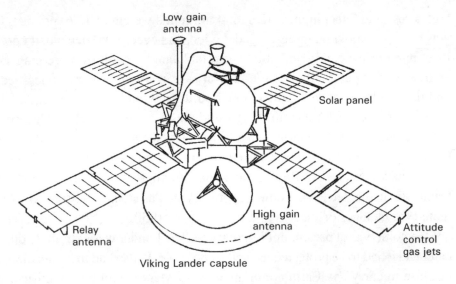

Low gain antenna

Solar panel

High gain antenna

Relay antenna

Attitude control gas jets

Viking Lander capsule

The Viking vehicle. (Vikings 1 and 2 were identical.)

be put into a closed path around Mars in much the same way as for Mariner 9. At the planned moment the Lander would be separated, and would descend through the Martian atmosphere to make a gentle descent on to the surface, leaving the Orbiter to its own devices. Not that the Orbiter was at the end of its career: far from it. Not only would it carry on with its own investigations, but it would also act as a relay to send back messages from the Lander. Indeed, without the Orbiter, the Lander would have been very restricted in its ability to communicate with Earth.

The launching vehicle – a Titan 3/Centaur rocket combination – had been well tested, and was expected to give no trouble. Neither was there any reason to doubt that there would be any real difficulty in putting the Viking into its path round Mars. The main risk was in the landing itself, which had to be completely automatic. Once the order to separate had been sent from Earth, the landing manœuvre could be neither stopped nor modified. Moreover, a time-lag would be inevitable, because radio waves would take over nineteen minutes to reach the Earth. When Viking 1 touched down, the distance between Mars and ourselves was 212,000,000 miles, while at the time when Viking 2 made its descent Mars was even further away.

In view of what they accomplished, these first two Vikings were amazingly small. The Orbiter was octagonal, 8 feet across, 10.8 feet high and a mere 32 feet across the full spread of its solar panels. The Lander was 10 feet across and 7 feet high, with a total unfuelled weight of 1270 pounds. It was a strange-looking, three-legged device, and everything depended upon its coming down gently and at an acceptable angle. A tilt of more than 19 degrees would put it out of action insofar as transmissions were concerned, and if the vehicle landed upon a large boulder it would be fatally damaged; the clearance of its base from the ground was a mere 8.7 inches. Unfortunately, boulders of 'dangerous' size were below the resolving limit of either Orbiter photography or Earth-based radar measurements, so that a good deal of luck was needed. Before the actual landings, it is fair to say that the scientists at the Jet Propulsion Laboratory at Pasadena (California) estimated the chances of success at no more than 50-50.

One point was of paramount importance. The Lander was extremely delicate, and it had to be protected as well as possible. It also had to be sterilized, because to carry any Earth contamination to Mars would be scientifically

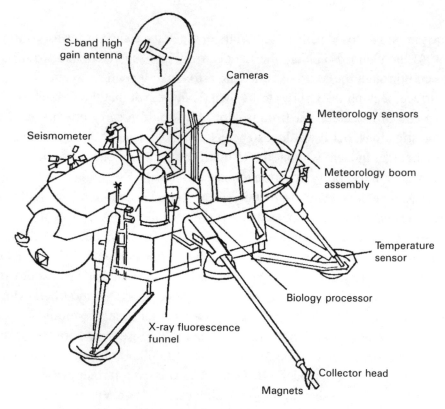

Viking Lander. (The biology box, gas chromatograph – mass spectrometer, X-ray fluorescence spectrometer and pressure sensor are internally mounted.)

disastrous. One way to kill any organism is by intense heat, and before being sent up the Lander was accordingly 'cooked' for forty hours in an oven heated to well above the boiling-point of water. There is every reason to hope that the treatment was effective. Of course, one can never be absolutely sure, but the risks have been eliminated as far as humanly possible, and no doubt this also applies to the Russian descent capsules.

Politics can never be kept out of science, unfortunately, and it is a fact that several of the most memorable space-shots have been timed for reasons that are not, strictly speaking, scientific. Thus the Space Age began on 4 October 1957, with a Russian satellite: the fortieth anniversary of the Bolshevik revolution. Luna 3, the first probe to go right round the Moon and send back pictures of the far side, was dispatched on 4 October 1959. With Viking, it was hoped to make the initial landing on 4 July 1976, American Independence Day, but at an early stage it became clear that this could not be done. There was trouble before the launching, and Viking 1 was delayed. Eventually the two vehicles were reversed, so that the probe which should have been Viking 2 became Viking 1 (and will so be called hereafter). To the great credit of all concerned, no attempt was made to speed things up. This might have jeopardized the whole mission, and the delay was accepted, though doubtless with a certain amount of regret. Actually, a further delay was encountered later on, because the original landing site had to be rejected, but by then the Independence Day target had been missed in any case.

Viking 1 finally took off on 20 August 1975, and Viking 2 followed on 9 September. I do not propose to say much about the launching itself, or about the journey through space; it is enough to note that there were no major hitches. On 19 June 1976 Viking 1 entered a closed orbit round Mars, and work began immediately. (By then both the Martian satellites had been photographed, but I propose to leave all discussion of these odd little worldlets until Chapter 14.)

Unlike Mariner 9, the Viking did not immediately set out to study the whole surface of Mars. Its first task was to check on the Chryse landing site in order to make sure – so far as possible! – that it really was a suitable place. But even before entering orbit, it had sent back some truly magnificent far-encounter pictures which surpassed anything that Mariner 9 had been able to do. The ochre tracts, the dark areas, the basins and the volcanoes stood out splendidly, and the improvements in photographic technique were obvious at once.

The pictures which came back soon after Viking had begun its main programme were absolutely staggering. One view, of the towering Olympus
Mons, was obtained on 31 July from a distance of 5000 miles; it was mid-
morning on Mars, and the volcano was wreathed in clouds which extended
up the flanks to an altitude of at least twelve miles. There could be little
doubt that these clouds were of water ice, condensed out of the atmosphere
as it cooled while moving up the slopes of the volcano. The giant caldera,
formed by the collapse of the volcano top, was splendidly shown above the
uppermost clouds. Another picture, taken later, showed the even larger
caldera of Arsia Mons, one of the other chief members of the Tharsis group.
Then there were smaller features such as the well-formed, 11-mile crater
now called Yuty, in the Chryse area not too far from the intended landing
site. Yuty is remarkably lunar in aspect, with regular, terraced walls and a
massive central peak crowned by a pit.

On 3 July Viking sent back a picture of part of the tremendous Valles
Marineris. From a range of 1240 miles, the Valley was seen to have destroyed part of the wall of a well-formed crater, and the effect was unlike
anything seen before. Remember, the Valley is several miles deep, and dwarfs
all terrestrial canyons. (On this picture there is a strange-looking black ring.
This ring is shown on many of the Viking pictures, but I hasten to add that it
is due to a slight defect in the optical system, and is in no way Martian.)
Then there were the fault zones, causing valleys which are widened partly
by collapse and partly by mass wasting, i.e. the downslope movements of
rocks due to gravity. Also of special interest were the subsidence features
which could well have been produced by the melting of ice below the surface; there were sand-dunes, such as the dune field near the north wall of the
Ganges Chasma; there was a canyon leading into the Valles Marineris complex; there were extensive lava-flows, and there were one or two amazing
pictures, one of which showed a mountain clump which, by chance lighting
and shape, gave an uncanny resemblance to a human face! Predictably, all
the usual eccentrics were excited by the 'face', and produced the usual stories about lost Martian civilizations, abandoned cities and the like; also predictably, it was claimed that The Authorities knew all about it, and were
doing their best to conceal the true facts. It is fair to say that 'conspiracy
theories' are the trademarks of cranks, and so it proved in this instance. The
'face' is an ordinary rock, as was finally proved in 1998 when it was imaged
from a different angle by Mars Global Surveyor.

Riverbeds were much in evidence, and by now there were very few scientists who believed that they could be anything other than dry watercourses. All in all, the Viking pictures added even more interest to what Mariner 9 had already told us, but the main objective was always to study Chryse, where the Lander was due to touch down.

Alarm signals had been sounded from Arecibo in Puerto Rico, where the immense radio telescope, built in a natural hollow in the ground, had been engaged in radar mapping of the Chryse area. There seemed to be a greater degree of roughness than had been expected, and the Orbiter pictures added to the general unease, because quite apart from the desirability of coming down at an angle of less than 19 degrees it was also essential to avoid landing upon a rock which would puncture the space-craft and put it out of action. Eventually the original site was abandoned, and a second target area selected, still in Chryse but rather to the north-west. This seemed to be better, but the final choice was still further west, and the die was cast. On 20 July, the great attempt began.

Absolute precision was impossible, and all that could be done was to aim the probe as accurately as possible and hope for the best. There was every prospect that it would land somewhere inside a restricted 'ellipse of uncertainty', and there were no detectable boulders within range, though the planners were unpleasantly conscious of the fact that neither the radar nor the Orbiter mapping could show features which would still be large enough to cause disaster.

The capsule that separated from the Orbiter was made up of three main sections: the Lander itself, a cone-shaped aeroshell made up of aluminium alloy, and a base cover. The Lander was enclosed, while the braking engines were part of the aeroshell and the parachute was in the cover. The whole procedure was different from that of a Moon landing, because the atmosphere of Mars, thin though it may be, is dense enough to cause definite friction, which involves heat. It also means that parachutes can be used, though they cannot cope with the full slowing-down process on their own.

When the descent capsule left the Orbiter, it began its gradual 'coast down' to Mars. It continued to do so, unhampered by any atmospheric drag, for more than three hours, but at an altitude of 800,000 feet above the surface – that is to say, around 150 miles – the atmosphere started to make its presence felt. In preparation for this, the capsule had been turned so that the aeroshell and its heat-shield faced the direction of travel. By now the cap-

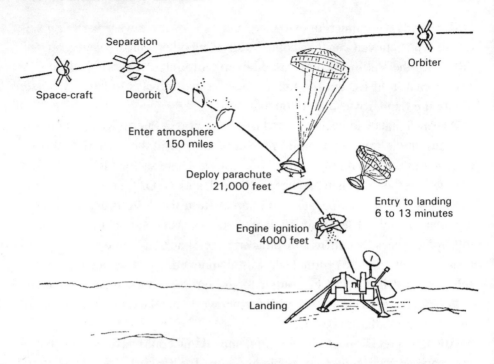

Descent sequence of the Viking Lander.

sule was moving at 10,000 m.p.h., which is not far short of three miles per second, and as it dropped lower and lower the frictional heating built up. The heat-shield proved adequate: the deceleration reached its maximum value at 15 to 18 miles above ground level, and for a brief period the path of the capsule levelled off into horizontal flight because of the aerodynamic lift provided by the capsule. With continued slowing-down, the heating became less violent, and the descent could be resumed. At 19,000 feet from the Martian surface the velocity was a mere 1000 m.p.h., and by now, of course, the capsule was well below the level of the tops of the Tharsis volcanoes. This was the moment for the parachute to be deployed, and seven seconds later the aeroshell, its work done, separated from the Lander and drifted away, to fall to the ground well away from the main site.

The last part of the descent was, obviously, the most critical of all, and the scientists back on Earth could know nothing about it, because of the 19¾-minute delay in the reception of signals. In fact the parachute worked perfectly, and within a minute the Lander's velocity had dropped so much that the effects of the winds on its horizontal travel were measurable. The legs were extended to the landing position, and when the radar altimeter gave the height at a mere 3900 feet the parachute was jettisoned. At the same moment the three terminal descent engines of the Lander itself came into play, and seconds later the 440,000,000-mile journey was over. The final touch-down speed was less than 6 m.p.h. Only an hour and a half later, scientists in the control centre at Pasadena were examining the first pictures ever to be successfully transmitted from the surface of Mars.

The touch-down position was a mere twenty miles from the planned impact point, well within the 'ellipse of uncertainty'. And yet luck had played its part. The Lander came down only 25 feet from a boulder which was ten feet across and three feet high – more than big enough to have caused fatal damage if the landing had been made on top of it. As soon as I saw the picture of it, I was reminded of the comment made in the television studio when we were demonstrating with models: 'If you drop them from a great height, they break!'

The very first picture, taken immediately after touch-down, showed that the entire landscape was strewn with rocks. One, near the centre of the picture, was four feet across, and six feet away from the Lander. One pad of the third leg was shown, and had penetrated only 1.4 inches into the soil, so that clearly the Martian surface was reassuringly solid, even though it was later found that another of the Lander's legs was covered with sand. The overall impression was of a barren, rocky desert, with extensive sand-dunes as well as pebbles and boulders.

What, then, about the colour? On the sol after the landing, Viking sent back a colour picture showing that the fine, granular material so much in evidence was rusty red, while most of the rocks were of similar hue, so that presumably the red material – probably limonite (hydrated ferric oxide) – formed a thin veneer over the darker bedrock. One major surprise was the colour of the sky. The first pictures showed it to be bluish, but this proved to be wrong; corrections were made, and the sky showed up as pink. Apparently the pinkness is caused by the innumerable fine dust-particles suspended in the atmosphere, and which absorb the blue light from the Sun. Later Viking pictures showed a sky which can best be described as 'yellowish pink'. The pioneer astronauts will find the scene unearthly in every sense of the term.

As the days (and sols) went by, more and more pictures came through, and each provided new topics for discussion. The colour pictures were naturally the most spectacular, though it must be stressed that these are actually computer reconstructions of three black-and-white images taken through filters. The main discussions centred round the rocks, which were seen to be of various types. Some were coarse-grained and knobbly, while others were lighter-coloured and irregularly shaped, and there were many blocks of considerable size. There was little doubt that the rocks were of volcanic origin. Superficial comparisons are always dangerous, but recently I looked first at a black and white picture from Chryse and then at a black and white photograph I had myself taken in Iceland some years ago. It was not easy to tell which was which. I also looked at a colour photograph of the Tibesti Desert, about a thousand miles south-west of Cairo; here the resemblance to Chryse was even more striking – redness and all.

All round the Lander there was abundant evidence of wind-blown material, and some of the rocks were obviously smoothed, but others were sharp, and this caused a certain feeling of surprise. Meanwhile, a great deal of extra information was coming through. Temperature measurements were made, and it was confirmed that Chryse is a distinctly chilly place. Even at the hottest time of the sol – around 2 p.m. local time – the temperature was no more than –22 degrees Fahrenheit, and just after dawn a thermometer would have registered at least 120 degrees below zero Fahrenheit. Remember, however, that Chryse is some way from the Martian equator.

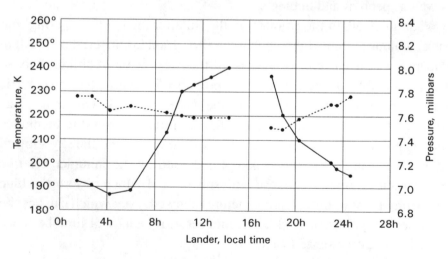

Temperature (degrees Kelvin) (continuous line) and pressure variations on the Chryse site (Viking 1) (dotted line) over a typical sol.

Winds were light. They were strongest at about 10 a.m. local time, but even then they were no more than 14 m.p.h. breezes; later in the sol they veered, dropping to around 4 m.p.h. and coming from the south-west rather than the east. The pattern seemed to be fairly regular, sol after sol. The pink sky was cloudless, and there was no evidence that material was being blown around in dust-devil fashion. During one of our BBC television programmes, Dr. Garry Hunt even made a Martian weather forecast, based upon information sent back from what he correctly called Man's most remote meteorological station. His forecast was: 'Fine and sunny; very cold; winds light and variable; further outlook similar.' Not surprisingly, he proved to be completely accurate!

One of the most important of all the investigations was the analysis of the atmosphere. In fact, this had already been started during the descent of the landing capsule. Temperatures recorded during the descent included +27 degrees Fahrenheit at a height of 130 miles, but this does not imply that the upper atmosphere is 'hot'; scientifically, temperature is measured by the rate at which the various atoms and molecules move around, and is not the same thing as everyday 'heat'. At 85 miles above the ground the temperature was found to be -216 degrees Fahrenheit. Nitrogen was detected, which pleased those scientists who were anxious to find life on the planet. It was also shown that, contrary to the data reported from Russia's Mars Fleet, there was very little argon.

Argon, incidentally, is of special importance. It comes in three varieties, known as Argon-36, Argon-38 and Argon-40 (the chemical abbreviation for argon is A). The first two types are due to the decay of radioactive potassium, while A-40 is probably 'left over' from the very early stage of a planet's existence, when it was still being formed out of the original solar nebula. The Earth's atmosphere contains three hundred times more A-40 than A-36, and five times as much A-36 as A-38. For Mars, it was found that A-40 accounted for roughly 1.6% of the total atmosphere, and that there was little of the other varieties of argon; the ratio of A-40 to A-36 was about 3100:1. From this, it would seem that Mars must have little radioactive potassium in its crust.

The latest analyses show that the make-up of the Martian atmosphere is: 95% carbon dioxide (CO_2), 2.7% nitrogen (N_2), 0.03% oxygen (O_2), 1.5% argon, and traces of other gases – such as krypton and xenon, both of which are also found in our own air, though admittedly they are in short supply. The ground pressure was the expected 7 millibars, as against an average of 960 millibars for the Earth's air at sea-level, but as the sols passed by it was

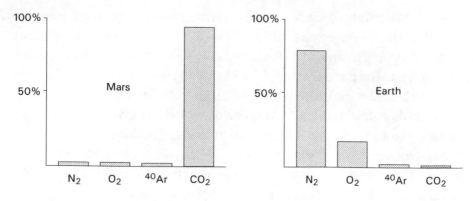

Relative compositions of the atmospheres of Mars and Earth.

found that the pressure was falling. The decrease was no more than about 0.012 millibar per sol, but it was definite, and there seemed to be a good explanation for it. Winter was approaching in the southern hemisphere, and the polar cap was starting to grow; the carbon dioxide condensing to form the southern cap was being withdrawn from the atmosphere in sufficient quantity to explain the falling pressure. Of course, the decline did not continue indefinitely; even before Mars passed out of range in November 1976 it had stopped. It is a purely seasonal phenomenon.

Water vapour was also detected, but here the main information was drawn from the Orbiter. The whole question of moisture is so vital in the search for life that I propose to defer discussion of it until the next chapter, together with the analyses of the surface materials carried out by the Lander.

All in all, Viking 1 performed almost faultlessly. The only disappointment was the failure of the seismometer, or Marsquake recorder, which would – it had been hoped – register considerable ground movement; after all, there are frequent tremors on the Moon, which would be expected to be far less active than Mars. Unfortunately, the cage which had protected the seismometer during the long flight refused to shift, and all efforts to persuade it to move proved abortive. Scientists could only hope that the seismometer of Viking 2 would work; and Viking 2 was already poised for descent. It had entered Martian orbit on 7 August, and was ready to dispatch its Lander to the chosen target.

As with its predecessor, there were qualms about the site. Cydonia seemed to be unpleasantly rough. Pictures taken from the first Orbiter were not encouraging, and there was no help from Arecibo, because Cydonia is so far north on Mars that it is out of the range of Earth-based radar. After much deliberation, it was decided to abandon Cydonia and substitute Utopia Planitia, a broad plain 4600 miles north-east of the Viking 1 site, which was expected to be flattish. And it was here, on 3 September, that the second Lander touched down. The position was given as latitude +48 degrees, longitude 226 degrees west.

Events did not go as smoothly as with the first mission. After the descent capsule separated, there was a momentary power failure in the Orbiter – slight, but enough to allow the space-craft to drift off its main communications link with Earth. Unless something could be done, all the information collected by the Lander during its descent would be lost, because the data were to be relayed immediately, and if the Orbiter could not pick up the signals all the priceless information would merely beam uselessly into space. Prompt action was called for. The planners were able to use the low-power communications link, and no data were permanently lost, though it was not until after touch-down in Utopia that the trouble was completely cured and the main communications link restored.

It was time for Mars to produce yet another of its seemingly never-ending surprises, and it duly did so. Instead of being a gently-rolling landscape, with sand-dunes and windblown dust, Utopia seemed to be of the same general aspect as Chryse. One investigator commented that it looked like 'a forest of rocks'. There were no major craters in sight – the closest large formation, Mie, lies 125 miles to the west – and there were no boulders as large as those in Chryse, while the rocks looked 'cleaner' and the general terrain was slightly flatter. On the other hand, medium and small rocks were abundant, and most of them were vesicular. Vesicles are porous holes, formed as molten rock cools at or near the surface of a lava-flow, so that internal gas-bubbles escape. There is absolutely no doubt that Utopia is a volcanic landscape, and this is borne out by various other rock-types on view. There are breccias (i.e. rocks made up of fragments of other rocks welded together under the influence of high pressure and heat) and there is at least one splendid example of a xenolith – that is to say, a 'rock within a rock', probably formed when a relatively small rock was caught in the path of a lava-flow and was coated with a molten envelope which subsequently solidified. There is also a great quantity of fine material similar to that in Site 1, but there are no comparable sand-dunes in sight, though there is a winding feature which looks suspiciously like the bed of a long-dry stream.

Otherwise, the results from the second Lander have not been significantly different from those of the first. The mean atmospheric pressure, 7.7 millibars, indicated an appreciable difference in level between Chryse and Utopia. Temperatures ranged between −23 and −114 degrees Fahrenheit. Winds remained light and variable, though with occasional gusts of over 30 m.p.h. The seismometer worked, to the relief of everyone, but no Marsquakes were recorded, which was a distinct surprise; if the instrument had been placed in, say, Southern California it would have recorded tremors constantly. It did, however, send back results of the tremors produced by the mechanism of the Lander itself when various operations were being carried out. The first Marsquake was recorded from Lander 2 in February 1977.

Meantime, some new manœuvres were in hand. It was known that for some weeks after 10 November there would be a loss of contact with Mars, because the planet would be too nearly behind the Sun. Superior conjunction was due on 27 November, and signals would be hopelessly blocked by the solar radio emission, after which nothing more could be expected until shortly after Christmas. Therefore, the Orbiters were ordered to do what may be called a 'general post', so that Orbiter 1 could act as the relay for Lander 2, releasing Orbiter 2 to carry out its independent programme. Lander 1 would go into what was termed its reduced phase, sending back less data every sol.

During the peak period, a Lander can contact Earth in two ways: either direct, or via the Orbiter as a relay. The quality of the transmission is the same, but the direct link can be used for less than one and a half hours per sol (a period known as 'real time'), whereas the relay link is capable of sending back much more information much more rapidly. (All commands sent from Earth *to* the Landers are direct.) The Orbiters had been put into paths which took them round Mars in exactly one sol, so that they came back over their relevant Landers at the same time each revolution, a situation which is known as being in synchronous orbit. On 11 September this was altered. Orbiter 1 fired its engine for sixteen seconds, which was enough to break the synchronous link with its Lander and reduce the orbital period to 21.3 hours. In other words, the Orbiter now reached its lowest point every 21.3 hours instead of once every sol. Since it reached its lowest point ahead of time, the previous day's reference point would lie to the west, by a distance equal to the distance that the planet's surface can rotate in 3.3 hours (the difference between a sol and the Orbiter's revised period). Therefore, to a Martian observer it would seem that the Orbiter would be further east each day, and would appear to be 'walking' round the planet.

By 24 September the lowest point in orbit was directly over Lander 2, and another manœuvre restored the period to its synchronous value, so that Lander 2 was again fully supplied with a relay vehicle. Orbiter 2 had begun a similar 'walk', and continued to send back excellent pictures as well as a vast amount of miscellaneous information. All was going well.

CHAPTER 11

The Search for Life

There must be rivers on Mars. The mere existence of continents and oceans
proves the action of forces of upheaval and of depression. There must be
volcanic eruptions and earthquakes, modelling and remodelling the crust.
Thus there must be mountains and hills, valleys and ravines, watersheds
and watercourses . . .

No, this is not a report from the Jet Propulsion Laboratory. It was written by
the British astronomer R. A. Proctor as long ago as 1871. It was then thought
that Mars must in many ways be very similar to the Earth, in which case life
there seemed only too likely; after all, why not?

We now know that the atmospheric pressure is too low for surface water
to exist, though the old riverbeds show that it used to do so in the distant
past. Ice is another matter, and the early Mariner findings that the polar caps
were made up of solid carbon dioxide came as a distinct disappointment.
Yet even before the first Mars flights it had been thought that the caps were
nothing more than wafer-thin layers of frost, containing only a very small
quantity of water locked in a frozen state. I once estimated that if all the ice
could be melted at once, the water produced would be insufficient to fill a
lake the size of Wales and the depth of Windermere – which shows how
wrong one can be, even when taking the best evidence available!

Probably the most important of all the discoveries made during the
early part of the Viking mission was that the caps are not made up of
pure solid carbon dioxide; at each pole there is a residual cap of water
ice, around 600 miles in diameter for the northern cap and over 200
miles for the southern. During its 22nd revolution round Mars, the Viking
2 Orbiter made some temperature measurements of the north polar regions which seemed to be conclusive. The dark area surrounding the cap
rose to a maximum temperature of –28 to –27 degrees Fahrenheit, while

the white cap itself was colder, at –90 degrees Fahrenheit. This is chilly enough, but it is well above the limit at which solid carbon dioxide could persist, and it followed that the residual cap – that is to say the part of the cap which is always present, and does not vanish during Martian summer – must be made up of H_2O ice, probably many tens of feet thick. The same was presumably true of the southern cap; each was overlaid by a seasonal layer of carbon dioxide ice.

The two caps are not alike, either in extent or in composition. This is because of the more extreme climate in the south; remember that southern summer falls when Mars is at perihelion. Predictably, the southern cap can extend down to latitude 55 degrees south, which is larger than its northern counterpart can ever be, though – as we have noted – the southern residual cap is much the smaller of the two when at its minimum size. It can shrink to a diameter of between 200 and 250 miles, whereas the northern cap is never much below 600 miles across. Moreover, the north residual cap is indeed made up of water ice, while the southern consists of a mixture of water ice and carbon dioxide ice. During summer the carbon dioxide coating of the northern cap disappears completely, but this is not so for the southern cap, where some carbon dioxide ice always persists. During northern spring and summer the Martian atmosphere is comparatively clear, and transparent, so that the carbon dioxide ice is exposed to solar radiation and dissipates, while during southern spring and summer the atmosphere is more dust-laden, protecting the upper carbon dioxide layer from the rays of the Sun.

While a cap is growing, clouds of carbon dioxide form over it, making it difficult for us to see just what is happening; by the time that the 'polar hood' disappears the cap is already well advanced in its seasonal cycle. Note, incidentally, that when a cap shrinks it does not 'melt'; it sublimes – that is to say, it turns directly from the solid to the gaseous state as the temperature rises. So far as atmospheric water vapour was concerned, Viking found that if all of it could be condensed it would cover Mars with a surface layer of nothing more than frost – whereas the release of water in the ice-caps would produce a layer at least 30 feet deep. All the same, the thinness of the atmosphere means that it can be saturated, with the greatest amounts of water vapour in low-lying areas such as Hellas. Ground ice tends to evaporate after sunrise, producing ground fogs, while during afternoon the vapour condenses once more as ice. According to the Viking measurements, the height of the Martian tropopause was given as around 75 miles, with carbon dioxide clouds rising to 15 miles and water ice clouds to 5 miles; dust-storms could attain at least 25 miles, though, as we have seen, the volcano-tops can often protrude above the uppermost layer of dust.

Once the Landers were safely down, they could begin to analyze the surface material, and see what elements were present there. Each Lander was equipped with a surface sampler, known more commonly as a grab, made up of a collector head on the end of an ingeniously-constructed retractable boom composed of two ribbons of stainless steel, welded together along the edges. The head itself was, basically, a scoop with a movable lid. It was capable not only of collecting Martian 'soil' and bringing it back inside the Lander for analysis, but also of digging a trench; it was even strong enough to overturn modest rocks and sample the soil beneath. Of course, all the Lander manœuvres were first simulated on Earth to make sure that they worked properly.

There was an initial alarm with the grab of Viking 1, because a latch-pin jammed and prevented the collecting operation from being completed. The planners decided to extend the boom beyond its first position, and this proved to be successful; the obstructing pin was released, and fell to the ground, where it was subsequently photographed. (It was ironical that a 3-inch pin almost wrecked one of the most complex and delicate experiments ever designed.) Finally, on the eighth sol after arrival – 28 July by our calendar – the grab secured a sample, and also dug a trench three inches wide, two inches deep and six inches long. It was significant that the sides of this miniature trench did not collapse, as would have happened if the material had been very soft.

The first studies of the composition of the Martian material were made by using a source of X-rays carried in the space-craft. When elements are bombarded with high-energy X-rays, the well-known phenomenon of fluorescence makes them emit X-rays of lower energy, and every element produces its own characteristic emissions, thereby betraying its identity. For once Mars provided no real shocks. It was found that there was 12% to 16% of iron, 15% to 30% of silicon, 3% to 8% of calcium, 2% to 7% of aluminium, and 0.25% to 1.5% of titanium, as well as various minor constituents. Comparisons with the Earth and Moon showed that the nearest analogy was the make-up of the lunar maria, or waterless seas. This was hardly surprising, since it had long been known that the Moon's maria are rock-strewn lava-plains.

Another interesting study concerned magnetic materials. Arrays of small magnets fixed to the body and sampler arms of Viking 1 were able to pick up numerous particles, and it was concluded that the soil contained 3.7 per cent of magnetic material, a large fraction of which was probably an iron oxide called magnetite. Only much later, with the Pathfinder and Global Surveyor probes of 1997–98, did it become possible to make more detailed studies of the surface materials, and to establish that Mars has a weak but definite overall magnetic field.

With a vast amount of locked-up water, an atmosphere containing a definite amount of nitrogen, and a temperature which was not impossibly low, the prospects of finding life appeared to be reasonably bright. Nothing more than very lowly organisms could be expected, but the discovery of life in any form would be immensely significant, and the experiments were put in hand as soon as it was practicable.

I do not propose to go into great detail, and the account given here is badly over-simplified, but at least I hope it will serve as a guide. Basically there were three experiments, all of which were tried by the Landers. They were:

(1) *The Pyrolytic Release Experiment* – Pyrolysis is the breaking-up of organic compounds by heat. The experiment was based upon the assump-

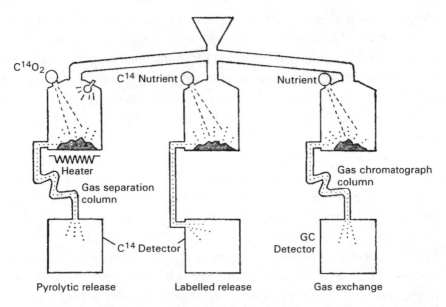

The experiments involving the Viking search for life on Mars.

tion that any life on Mars would contain carbon, as is the case on Earth. One species of carbon, known as carbon-14, is radioactive, so that when it is present it is relatively easy to detect. The scheme was to treat the Martian samples with this convenient carbon, and see whether they assimilated any of it.

The sample was heated in the test chamber for five days. The atmosphere in the chamber was similar to that on Mars, except that carbon dioxide and carbon monoxide, duly labelled with carbon-14, replaced part of the natural atmospheric gases. To make the situation even more realistic, the chamber was illuminated by artificial sunlight, to help in the phenomenon of photo-synthesis (the production of organic compounds by carbon dioxide and water by green plants). The only major departure from actual conditions was that the artificial sunlight lacked the short-wave ultra-violet radiations which come from the real Sun, and which might well have damaged any Martian organisms present.

After an incubation period of eleven days, the chamber was heated to 1160 degrees Fahrenheit, hot enough to break up any organic compounds (pyrolysis, as we have noted). A stream of helium gas, which is inert, then flushed out the chamber, and swept the vaporized pyrolysis products into a detector capable of identifying the carbon-14 taken up by the Martian organ-isms – if any. Should no carbon-14 be found, it would be safe to assume that there was no biological activity or, in other words, no life; but a positive result would not be conclusive, because a 'first peak' of radioactivity in the vaporized gases could also be caused by chemical processes which did not involve living organisms. A second heating, this time to 1290 degrees Fahr-enheit, should convert any trapped organic compounds into carbon dioxide, and again the labelled carbon-14 should show up, producing a second 'radio-active peak'.

Two peaks were shown, but neither was at all conclusive, and the same pattern was followed in later experiments from both the Landers. This in-cluded samples which had been deliberately sterilized, so as to kill any organ-isms which might be present.

(2) *The Labelled Release Experiment* – This also involved the useful car-bon-14, and assumed that the addition of water to a Martian sample would trigger off biological processes if any organisms were present. Again the atmosphere provided was similar to that of Mars, though in the first experi-

ments the chamber was kept dark during the incubation period. The sample brought in by the grab was moistened with a nutrient which contained carbon-14 as well as various other substances which are taken in by Earth organisms, while the atmosphere above the sample was kept under close surveillance by a set of detectors capable of showing any traces of radioactivity. Organisms would be expected to give off gas containing carbon, and the carbon-14 would therefore betray it. Therefore, if any radioactivity in the atmosphere of the chamber were found, it would be a strong pointer to the existence of life.

The results of the preliminary experiment were startling. As soon as the first drops of nutrient were put into the chamber, the level of radioactivity rose sharply, and the excitement at Pasadena was considerable. Alas, things did not turn out to be so straightforward as had been hoped. Within a sol, the level of radioactivity dropped, and after a week it had levelled out. At the end of a seven-day period a few more drops of nutrient were put into the chamber. The immediate result was the release of more labelled carbon-14, but subsequently the total amount actually dropped by 30 per cent, after which it rose again very slowly.

The investigators simply did not know what to make of it. The behaviour of the sample did not indicate either 'life' or 'no life'; it was a complete puzzle. In a later test, a sample was 'cold-sterilized' in a way which should have made it possible to distinguish purely chemical reactions from the more sensitive biological ones. The values obtained were lower than for untreated samples, but higher than those for fully sterilized ones.

(3) *The Gas Exchange Experiment* – This was expected to be the most sensitive of all the tests, and for once carbon-14 was not involved. The main assumption was that any biological activity on Mars would involve the presence of water, and the idea was to see whether providing a sample with suitable nutrient would persuade any organisms to release gases, thereby altering the composition of the artificial atmosphere inside the test chamber. This atmosphere was made up of carbon dioxide, together with the inert gas krypton to act as a calibration standard, and helium to bring the pressure up to an acceptable level. The sample was held in a porous cup above the chamber floor, and a rich mixture of common nutrients in water solution was added to the bottom of the chamber – a mixture which was generally nicknamed 'chicken soup', though I doubt whether any gourmet would have

found it palatable! For the first week the level of liquid was kept below the cup, so that only water vapour could be transferred to the sample. After that, the level of the liquid was raised so that the sample was actually wetted. The atmosphere in the chamber was regularly tested to see if any changes were occurring.

Again there was an immediate surprise. As soon as the so-called chicken soup was put into the chamber, there was a violent release of oxygen from the sample – fifteen times as great as could be explained by any known process. Carbon dioxide was also released, together with a certain amount of nitrogen, but after a while there was a noticeable levelling-off, and when the soil was wetted no more oxygen was set free; the carbon dioxide content decreased. The investigators more than two hundred million miles away frowned thoughtfully. Nothing seemed to make sense, and even the fact that the sample came from a desert, thereby being unused to appreciable moisture, did not help much in finding an explanation. Eventually they came round to thinking that the reactions were chemical rather than biological, due either to the release of oxygen stored in the sample or else by the reaction of some unstable oxidant with the nutrient.

I have dwelt at some length upon these experiments because they were so important and so novel. Trenches were dug all round the two spacecraft, and as many samples as possible were examined (it was commented that the areas round the Landers began to look rather like sites for mining operations) and samples were also obtained from beneath rocks which were overturned or pushed aside by the grabs. One rock, tackled by Viking 2 in October, obstinately refused to budge, so that evidently most of it lay buried beneath the surface, but a smaller one was pushed a total of eight inches, and on 25 October, at 6.45 a.m. Martian time, a scoopful of soil was obtained from a third rock which had been shifted by the grab half an hour earlier. Yet with these samples, too, the results remained inconclusive. The situation was summed up very neatly by Dr. Gerald Soffen, one of the principal investigators, after some of the results from Lander 2: 'All the signs suggest that life exists on Mars,' he said wryly, 'but we can't find any bodies.'

I think that most people, including myself, expected the Landers to give us a final answer to the question of whether there is, or ever has been, life on Mars; but this did not happen, and even now, more than two decades later, we still do not know. But at least the Vikings achieved all that could have been expected of them, and all the experiments worked well. The Orbiters are still circling Mars, and the Landers are on the surface awaiting collection. Certainly they must be regarded as among the most successful of all space-vehicles of the twentieth century.

CHAPTER 12

Pieces of Mars?

In January 1998 I found myself in Antarctica, standing on the frozen ground and admiring the icebergs and the penguins. It is an amazing place – a continent larger than Europe, but with an indigenous population of zero. By now there are scientific bases there, and a major observatory is actually being built on the South Pole itself, but until recently Antarctica has been more or less undisturbed. This means that it is virgin territory so far as science is concerned, and in particular it is a happy hunting-ground for collectors of meteorites.

Most people are familiar with meteors or shooting-stars, which are débris left behind by comets. The average meteor is smaller than a grain of sand; if it dashes into the upper atmosphere, moving at anything up to 45 miles per second, it rubs against the air-particles, and sets up so much heat by friction that it burns away in the streak of radiation which we call a shooting-star. Meteorites are quite different, and are not associated with comets. It is believed that most of them come from the asteroid belt, and indeed there may be no difference between a small asteroid and a large meteorite – though the term 'meteorite' is conventionally restricted to an object which has actually landed on the surface of the Earth. Most of them are stones, irons or a mixture. Many thousands have been found, and some of them are large; the meteorite at Hoba West, in Southern Africa, weighs over 60 tons, and is still lying where it fell in prehistoric times.

Most meteorites are not recovered, because they are not identified, and to most people look exactly like ordinary pieces of rock, but things are different in Antarctica, where a meteorite may lie placidly for a very long period before any member of an expedition notices it. Careful searches have led to the recovery of many specimens, and one of these, discovered in 1984, was found to be of special interest when it was properly studied some years later. It is know as ALH 84001, because it was collected in the region of

Antarctica's icy Allan Hills. It is shaped rather like a potato; it is 6 inches
long by 4 inches by 3 inches, and it weighs over 4 pounds. When picked up
it was partly covered with black or fusion glass – rather as though it had
been dipped in tar. At the time there seemed nothing unusual about it, but
when it was examined in 1993 it was found to be what is known as an SNC
meteorite – and, it was claimed, came from Mars.

Twelve of these SNC meteorites are known. The first three to be identified came from Shergotty in India, Nakhla in Egypt and Chassigny in France
– hence the name. It may be of interest to give a list:

Name	Location found	Date found	Mass, grams
Chassigny	Chassigny, France	3 Oct. 1815	±4000
Shergotty	Shergotty, India	25 Aug. 1865	±5000
Nakhla	Nakhla, Egypt	28 June 1911	±40,000
Lafayette	Lafayette, Indiana	1931	±800
Governador Valadares	Governador Valadares, Brazil	1958	158
Zagami	Zagami, Nigeria	3 Oct. 1962	±18,000
ALHA 77005	Allan Hills, Antarctica	19 Dec. 1977	482
Yamato 793605	Yamato Mtns, Antarctica	1979	16
EETA 79001	Elephant Moraine, Antarctica	13 Jan. 1980	7900
ALH 84001	Allan Hills, Antarctica	27 Dec. 1984	1940
LEW 88516	Lewis Cliff, Antarctica	22 Dec. 1988	13
QUE 94201	Queen Elizabeth Range, Antarctica	16 Dec. 1994	12

The Nakhla fall, incidentally, was decidedly unusual. About forty stones
plummeted down from the sky, preceded by a cloud together with violent
explosions which alarmed the local residents. The total weight of the stones
was of the order of 88 pounds, and one of them landed on top of a dog,
which was unlucky enough to be at the wrong place at the wrong time. The
largest single SNC meteorite is the Zagami, which fell in Nigeria on 3 October 1962, and weighs 40 pounds. It landed a mere ten feet away from a
farmer who was trying to chase crows away from his cornfield; there was a
loud bang, a puff of smoke, a thud, and the meteorite buried itself in a hole
over two feet deep. It must have been a frightening experience!

Many of the investigators are confident that the SNC meteorites really do come from Mars. It is said that their composition is the same as that of the Martian surface, and that tiny quantities of gases trapped inside them are similar to the gases in the Martian atmosphere. They are igneous rocks, which crystallized out of molten lava in the crust of the parent body – in the case of ALH 84001 around 4.5 thousand million years ago, although apparently the other SNC objects are much younger. It all seemed convincing enough, though at the time when ALH 84001 was identified as belonging to the SNC type we did not have the advantage of the more detailed analyses carried out on Mars by the Sojourner probe over a year later, and the match does not now seem so perfect as it originally did.

What really brought ALH 84001 to public attention was the claim that it contained tiny features which were organic in nature – very small 'fossils', in fact. And on 6 August 1996 the Administrator of NASA, Daniel S. Goldin, issued a statement which is worth quoting: 'NASA had made a startling discovery that points to the possibility that a primitive form of microscopic life may have existed on Mars more than three billion* years ago. The research is based on a sophisticated examination of an ancient Martian meteorite that landed on Earth some 13,000 years ago. The evidence is exciting, even compelling, but not conclusive. It is a discovery that demands further scientific investigation. NASA is ready to assist the process of rigorous scientific investigation and lively scientific debate which will follow this discovery.' However, he added prudently: 'I want everyone to understand that we are not talking about "little green men". These are extremely small, single-cell structures that somewhat resemble bacteria on Earth. There is no evidence or suggestion that higher forms of life have ever existed on Mars'

President Clinton hailed the discovery as 'awe-inspiring', and the interest was by no means confined to the United States. It is interesting to look at some of the headlines in British national papers over the next week or so:

Fossil Clue to Life on Mars (*Daily Mail*)
Earth Welcomes Martian Invasion (*The Sunday Times*)
Mars Find Puts Life Back into Space Race (*Independent*)
No Longer Alone? Red letter day for science (*Evening Standard*)
It's Rock-Solid Evidence of Life on Mars (*Express*)

---

* American billion – three thousand million.

Life on Mars: Official! (*Daily Mirror*)
Clinton Hails Discovery of Life on Mars (*The Times*)

Yet from the outset, doubts began to creep in. In particular, the 'structures' were perplexingly small. Even the largest of them was no more that 500 nanometres long – and a nanometre is one-thousand-millionth of a metre, or less than one-hundredth the width of a human hair; a hundred times smaller than any micro-organism found on Earth. In shape it was said that they looked rather like worms. They might have been 'fossilized bacteria', but equally possibly they might have been inorganic. There were also oily molecules known as PAHs (polycyclic aromatic hydrocarbons), which can be produced by the decomposition of living organisms, but can also be produced by inorganic processes. It was significant that PAHs were also identified in Antarctic ice and in meteorites which were not suspected of having come from Mars.

Moreover, the chance of Earth contamination had also to be considered, and there was a parallel here with the case of the last-century Orgueil Meteorite, which had been said to show 'organized elements' indicative of past life; subsequently it was found that these 'organized elements' were due to ordinary tree pollen which had been collected well after the meteorite landed. All in all, the presence of past Martian life in ALH 84001 rested on very uncertain evidence indeed. As time went by, it was said that the biological explanation became less and less plausible.

Next, let us see whether the claim that the SNC meteorites are of Martian origin rests on a firmer foundation. How could they have reached the Earth – and by what process?

There is absolutely no chance that they could have been shot out of erupting volcanoes. Remember, the escape velocity of Mars is over 3 miles per second, and no volcanic outburst could possibly accelerate a rock to this speed. So the only possible answer is that the meteorites could have been blasted away from Mars by a violent impact.

Major impacts on Earth have happened in the past, and no doubt on Mars too. Go to Arizona, and some way off the famous Highway 99 you will find Meteor Crater, which is getting on for a mile wide and is a noted tourist attraction.* Another well-formed impact crater is Wolf

---

* Technically this name is wrong. It really should be Meteor*ite* Crater.

Creek, in Australia. There is even a popular theory that a colossal impact, 65,000,000 years ago, threw so much débris into the atmosphere that the Earth's whole climate changed, with disastrous results for the dinosaurs. (I must admit to being somewhat sceptical about this idea, but it has met with wide support, and certainly it cannot be ruled out.) If an impactor more than around a mile wide did hit Mars, some rocks might well be thrown clear.

Following up this principle, let us trace the career of ALH 84001. It was blasted away from Mars about 16,000,000 years ago; it could not have been put initially into an Earth-crossing path, and instead it entered an orbit round the Sun. At first this orbit was not very different from that of Mars itself, and there were various encounters, each of which weakened the gravitational link between the two bodies. Finally they separated completely, and ALH 84001 continued its journey alone; it had in fact become a tiny asteroid. By sheer chance its orbit eventually intersected that of the Earth, and around 13,000 years ago it plumped down in Antarctica, where it remained until the meteorite collectors found it.

All this is possible, but to my mind it all seems rather too glib. There is always a danger of jumping to conclusions, particularly when they sound both exciting and attractive.

Similar claims have also been made for another SNC meteorite, EETA 79001, which turned up in Elephant Moraine in Antarctica in 1980; it weighs 17.4 pounds, and is the second largest single 'Martian' meteorite ever found (only the Zagami Meteorite is larger). EETA 79001 had been said to contain microfossils, though these are even less definite than those of ALH 84001. One interesting point is that EETA 79001 is much the younger of the two; it seems to be a mere 180 million years old. If it were of Martian origin, it would have been blasted away from the planet 600,000 years ago. This would indeed be significant. If life on Mars still existed little more than half a million years in the past, it might still exist now.

Arguments continue; the jury is still out, and all I can do here is to give my own personal view. I admit that I do not believe that the wormlike structures in the SNC meteorites are true fossils, and I am also dubious about the Martian origin of any of the meteorites. But I may be totally wrong, and within a few years we ought to know. Before 2005, if all goes well, we will be able to send a robot probe to Mars, land there, collect samples and bring them home for analysis. This will settle the matter once and for all.

I await the result with the greatest possible interest. If I am right, I will not hesitate to draw attention to the fact. If I am wrong, then I will apologize suitably, withdraw my sceptical comments, and try to forget that I ever made them!

CHAPTER 13

Return to the Red Planet

Since the end of the Viking mission there have been six probes to Mars. Of these, two were aimed chiefly at the larger of the two satellites, Phobos, and I would prefer to postpone discussion of them until Chapter 14. Of the others, two were successful and two were not.

The first of these was America's Mars Observer, launched from Cape Canaveral on 25 September 1992 by a Titan 3 rocket. A great deal was expected of it, and a tremendous amount of planning and preparation had been undertaken (the total cost, incidentally, was 845 million dollars, which seems a great deal until you compare it with the cost of a nuclear submarine). It was about the size of a large office desk, and was crammed with the latest and most sophisticated equipment. There were eight experiments in all, including a magnetometer, a gamma-ray spectrometer and a CCD camera. The camera was expected to show details less than five feet across, and when using a fisheye lens would survey the entire planet once in every sol down to a resolution of below 1000 feet – so that, for instance, the Martian weather system could be followed. When the probe passed behind Mars, occultation experiments could be carried out to help in refining the analysis of the planet's atmosphere.

The launch was temporarily delayed by a hurricane in the Cape area, but when it eventually took place everything seemed to go according to plan, and Mars Observer was safely on its way. It cruised across space toward Mars, and prepared to enter an elliptical orbit round the planet. Its path would take it over both the poles, and would then be modified, ending up circular at a height of 233 miles; this would be 'sun-synchronous', so that Observer would always be above the daylight hemisphere of Mars. The planned active lifetime was a full Martian year.

The first image was obtained on 26 July 1993, when Observer was a mere three days' travel from its target; the picture showed an almost cloudless half-Mars. But then disaster struck. As part of the 'braking' manœuvre the radio transmitter was turned off, so that the fuel tanks could be pressurized and made ready to put Observer into its final orbit. Contact was never regained. The last signal came back on 21 August; after that – nothing.

Various contingency plans had been made. Fort example, if Observer received no commands from Earth for five days, it should have re-oriented itself with respect to the Sun and then switched to its backup receiver, but this did not happen. Frantic efforts to persuade Observer to call back met with no response, and at last, at the end of September, the NASA team had to admit defeat. Mars Observer was lost.

What had happened? We will never know. There may have been an on-board explosion, but it is equally possible that Mars Observer is still in orbit, either round Mars or round the Sun. Sadly, the chances of our re-locating it at any time in the future seem to be nil.

Next came Mars 96, launched on 16 November 1996 by Russia (no longer the U.S.S.R.). This was certainly the most elaborate Mars probe to date. It weighed 6180 kilograms, and was three-axis stabilized; it carried a total of twenty experiments, covering virtually all fields of research. The main scientific instruments were mounted in two articulated platforms, with the rest spread all over the body.

The vehicle was expected to reach Mars on 12 September 1997, but before that two small stations were to be dispatched and sent down to the surface. They would carry heat-shields to protect them when they entered the Martian atmosphere, and would then be slowed down by parachute; on impact they would be cushioned by airbags. One vehicle was aimed at Arcadia, the other at Amazonis. After arrival, four clamshell petals would open, unveiling the instruments so that they could begin their work. Subsequently two more landers would be dispatched, one to Acidalia and the other to Utopia. Each would carry a penetrator which would sink to nearly twenty feet into Mars, rather in the manner of a harpoon.

The ground stations were scheduled to last for at least a Martian year, while the Orbiter would continue mapping and undertaking gravitational studies as well as acting as a relay. Though the mission was Russian, the NASA teams had agreed to help, particularly in regard to navigation. Certainly the whole project was ambitious, and it was said that even if only half the experiments worked Mars 96 would still be a real triumph.

In fact none of them worked, because the space-craft never managed to break free from Earth's gravity. The fourth stage of the Proton launcher put Mars 96 into a parking orbit, but then obstinately refused to re-start to send the probe on its way to Mars. Then the space-craft itself separated from the launcher, and fired its own motors; alas, these were not powerful enough to achieve escape velocity, and a few hours later Mars 96 re-entered the lower atmosphere and burned away. Parts of it may well have landed in Bolivia or Chile, while the upper stage of the Proton re-entered over the Pacific on 17 November – causing some confusion, as it was at first believed to be the probe itself.

It was all very unfortunate; the Russians' ill luck with Mars continued. Everything now depended on the two U.S. probes, Mars Global Surveyor and Pathfinder. Global Surveyor was actually on its way at the time of the Russian disaster; it had been launched from Cape Canaveral on 7 November 1996. Pathfinder followed on 4 December, but travelled in a more economical path, and was the first to arrive by a matter of two months.

Pathfinder marked a new approach by NASA, and undoubtedly owed a great deal to what had been worked out by the Russians. It was not nearly so large or so massive as the ill-fated Mars 96; the cruise vehicle was comparatively small, with a diameter of 8.5 feet and a height of 5 feet. The lander was shaped like a pyramid, with a base and three triangular sides; it was of course computer-controlled, and was equipped with solar panels. The main camera was set atop a mast, standing 5 feet above ground level. The power which it could transmit was 160 watts; the active life was expected to be thirty days – which proved to be unduly pessimistic.

Pathfinder was launched from Cape Canaveral by a Delta-2 rocket, and there were no problems; neither were there any alarms during the cruise, which sent the vehicle into a long, gentle curve. By the time Pathfinder reached Mars it had covered a total of 309 million miles, and had been almost half-way round the Sun. On arrival the Earth was 119 million miles away. This is much greater than the distance between the Earth and the Sun, so that a signal from the space-craft would take over ten and a half minutes to reach Mission Control.

This time there was no attempt to put the space-craft into an orbit round Mars, and neither was there to be any attempt to bring it down as gently as with the Vikings. Pathfinder would crash through the Martian atmosphere, land, and bounce several times before coming to rest. To make this possible,

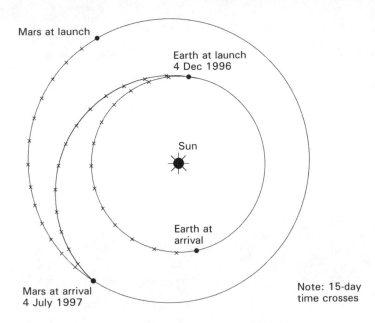

Mars at launch

Earth at launch
4 Dec 1996

Sun

Earth at
arrival

Mars at arrival
4 July 1997

Note: 15-day
time crosses

The Earth–Mars trajectory of Mars Pathfinder.

the whole vehicle was surrounded by airbags which would, it was hoped, provide enough cushioning to prevent any damage. Bearing in mind that space-craft are delicate things, the whole procedure seemed hazardous, to put it mildly, and some sceptics even saw an element of farce in it; Pathfinder was likened to a beach ball, and one writer wondered what any watching Martians would make of it. But the die was cast, and on 4 July 1997 – America's Independence Day – Pathfinder touched down on the Red Planet. The lander was promptly renamed the Carl Sagan Memorial Station, in honour of the astronomer who had so consistently championed the cause of exploring the Solar System by means of rockets.

Tension was high at Mission Control as the critical moment approached; of course, because of the ten-minute delay, the result – success or failure – would not be known until the whole manœuvre had been completed. Thirty-seven minutes before the planned landing time, Pathfinder was hurtling along at 16,000 m.p.h. relative to Mars, and was still 8000 miles away from the surface. The cruise stage was jettisoned, and final preparations were made for the landing. It would take place in darkness; it was night over that part of Mars.

Just over half an hour after leaving the cruise stage, Pathfinder hit the top of the denser part of the planet's atmosphere, and friction at once began to make itself evident. Pathfinder was now no more than a hundred miles above

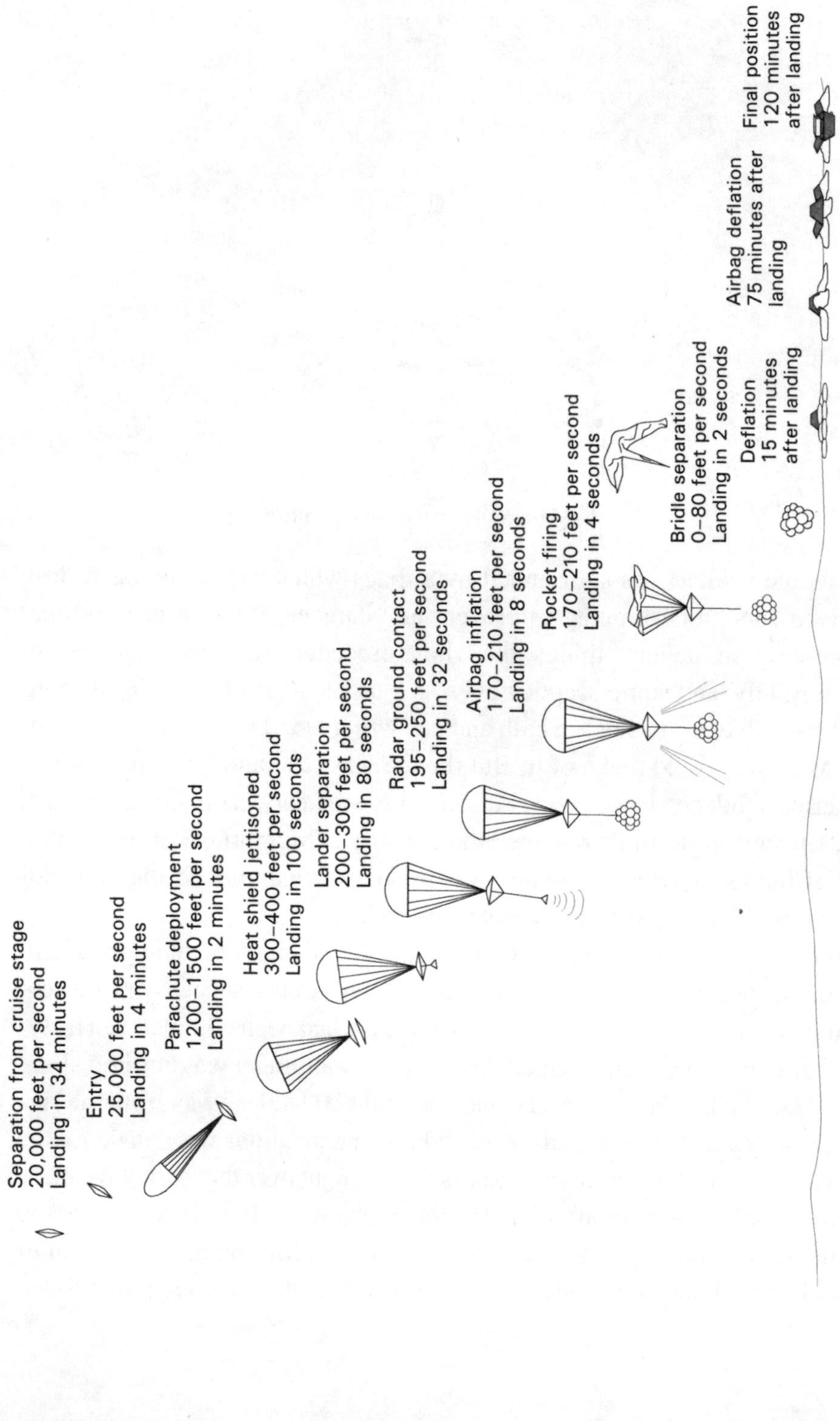

Separation from cruise stage
20,000 feet per second
Landing in 34 minutes

Entry
25,000 feet per second
Landing in 4 minutes

Parachute deployment
1200–1500 feet per second
Landing in 2 minutes

Heat shield jettisoned
300–400 feet per second
Landing in 100 seconds

Lander separation
200–300 feet per second
Landing in 80 seconds

Radar ground contact
195–250 feet per second
Landing in 32 seconds

Airbag inflation
170–210 feet per second
Landing in 8 seconds

Rocket firing
170–210 feet per second
Landing in 4 seconds

Bridle separation
0–80 feet per second
Landing in 2 seconds

Deflation
15 minutes
after landing

Airbag deflation
75 minutes after
landing

Final position
120 minutes
after landing

The entry, descent and landing of Mars Pathfinder.

the ground, and this was when the heat-shield played a vital rôle; it kept the temperature down to a tolerable level until the distance from the surface was no more than six miles, when the parachute system was deployed. The heat-shield had done its work, and was jettisoned twenty seconds later. The parachute took over the main task of slowing Pathfinder down, and it worked perfectly; after another twenty seconds the lander was lowered on the end of a long tether, with the speed now down to little more than two hundred feet per second.

At a thousand feet above the surface, eight seconds before landing, a radar command inflated the airbags, and four seconds later three solid-propellant rocket motors were fired, reducing the speed still further. Then the tether and parachute were jettisoned – otherwise they might have landed on top of the space-craft, causing a great deal of trouble – and Pathfinder literally fell the last hundred feet, impacting at 55 m.p.h. As planned, it bounced – not once, but at least fifteen times, possibly seventeen. The first bounce took it up well over five hundred feet, but the rest were less violent, and at last Pathfinder came to a halt, rolling along for about a minute after its final touch-down. Altogether it had bounced along for over half a mile, but it came to rest upright, so that there was no need for an extra manœuvre to allow Sojourner to emerge. Almost at once a signal came through, and the planners at NASA knew that their scheme had worked. It was certainly a moment for jubilation!

There was one minor hitch; an airbag had come down over the lander, and had to be shifted, but the problem was soon solved, and early on Sol 2 – the second day on Mars – Sojourner came out of the lander, crawled down the rear ramp, and was ready to begin its programme. All in all, it would have been difficult to imagine a more 'textbook' operation. It certainly compensated for the loss of Mars Observer.

Sojourner was a real midget, 26 inches long by 18 inches wide and with a height of less than a foot, which makes it about the size of an ordinary television set. It was six-wheeled, and was of course computer-controlled; it was equipped with solar panels which could keep it operational for several hours per sol. It was not exactly a quick mover, since its top speed was a modest 0.4 of an inch per second, but it was undoubtedly versatile, and could climb over small rocks; larger obstacles could be avoided. Naturally, it sent back its data by using the lander as a relay. When it was over thirty feet from the main station it could not be 'seen', but it was equipped with its

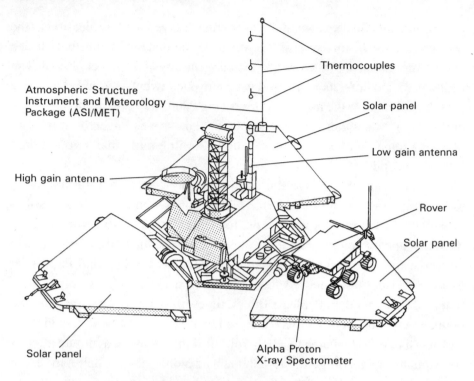

The Mars Pathfinder lander.

own cameras, so that NASA scientists at Mission Control, so many millions of miles away, could plan its path. It may have looked like a toy, but it was an amazingly sophisticated vehicle, and nothing like it could have been built at the time of the Vikings. Its estimated operational lifetime was no more than a week.

The site chosen for the landing was the region of Ares Vallis, at latitude 19.33 degrees north, longitude 33.55 degrees west. There were very good reasons for this, and a great deal of thought had gone into the selection; the area was reasonably flat by Martian standards, and was roughly the size of Cornwall.

I have already said something about the old riverbeds, which were certainly cut by running water. I have also made the point that if they were really ancient – that is to say, thousands of millions of years old – they ought to show more erosion than they actually do. In any case, there can be little doubt that they were once raging torrents, bringing rocks of all kinds down into the lower-lying areas. Area Vallis is one of these former rivers, lying north of Margaritifer Terra and south of Acidalia Planitia; there are others in the same general area, including the 680-mile-long Kasei Vallis. These debouch into what was expected to be an old flood-plain, in which case there ought to be a wide variety of rock types, so that Ares Vallis could well be what geologists call a 'grab region'. There would be volcanic rocks, and also rocks which had been hurled out of impact craters. The area, about 530 miles from the silent Viking 1 lander, was as accessible as any part of Mars.

It soon became clear that the predictions had been right, and that Pathfinder had indeed landed in what had once been a flood-plain. What seems to have happened is that volcanic activity below the Martian crust had melted ice which had accumulated in underground reservoirs, with the result that the crust collapsed and water gushed out to inundate the whole area. In fact, there was a temporary sea, and only gradually did the water disappear, leaving the area as barren and arid as it is today. No doubt the last rain on Mars fell tens of millions of years ago, or even longer.

The camera on the Sagan station was able to survey the whole area, and send back a panoramic picture. It was a fascinating scene. There were rocks of all shapes and sizes, there were dunes and mounds, and in the distance there were two elevations which were nicknamed Twin Peaks, though in fact they were not really very high. It was decidedly chilly. During the day-

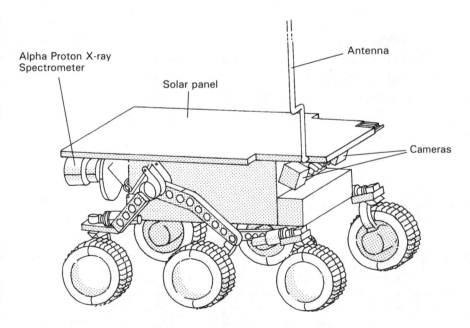

Alpha Proton X-ray
Spectrometer

Antenna

Solar panel

Cameras

The Sojourner rover from Mars Pathfinder.

time the temperature rose above freezing point, and there was one balmy day reaching 14 degrees Fahrenheit, but the nights were bitter, with temperatures well below −100 degrees Fahrenheit. The average atmospheric pressure was 6.74 millibars, which is not enough to allow liquid water to exist. Wind-speeds were fairly gentle, reaching 20 m.p.h. by night, and there were frequent 'dust devils', but in that thin air the wind had very little force, and was certainly too weak to damage the Station. Ice-crystal clouds were seen in the sky, probably at least ten miles up and corresponding to our own cirrus.

As Sojourner moved around it left a track in the soil, exposing darker material below; the soil itself was finer than talcum powder, and was likened to the fine-grained silt of the kind found in Nebraska in the United States. But the main interest lay in analyzing the rocks, and this was done by using an instrument known as the Alpha Proton X-ray Spectrometer, or APXS for short.

APXS carried a small quantity of the highly radioactive element curium-244, which emits alpha particles; these are really helium nuclei. Sojourner was able to go right up to the target rock and 'nuzzle' against it. The alpha particles sent out from APXS then bombarded the rock. In some cases these particles interacted with the rock and bounced back, while in other cases protons or x-rays were generated. The backscattered alpha particles, protons and x-rays were counted, and their energies measured. The numbers of particles counted at each energy level gave a clue to the abundance of the various elements in the rocks, and also to the rock types. All this may sound highly complicated, but it meant that APXS could carry out a detailed survey of the chemistry of the rocks, and this was what the NASA scientists really wanted to find out.

The rock nearest to the main station was the first to be examined by Sojourner; it was nicknamed Barnacle Bill, because of its outward appearance.* To the surprise of some of the geologists it proved to be similar to terrestrial rocks known as andesites, which are volcanic; another nearby rock, Yogi (so named because of a profile which was said to resemble Yogi Bear), was different, since it was basaltic and was less rich in silicon. Nei-

---

* Nicknames were distinctive; other rocks were known as Hassock, Sausage, Mermaid, Squash, Otter, Boo Boo, Picnic Basket, Flat Top, Moe, Wedge, Ginger, Stripe, Cradle, Soufflé, Baker's Bench and Desert Princess. Nobody can accuse the NASA planners of being lacking in imagination!

ther the rocks nor the soil bore much resemblance in composition to the 'Martian meteorites' found in Antarctica, though they were similar to the materials which had been found at the Viking sites. Rounded pebbles and cobbles were seen not only on the ground, but also in hollows in some of the rocks, which again suggested conglomerates formed in running water. There could no longer be any serious doubt that Mars had once been a watery world, and this inevitably raised the question of whether it had ever supported life.

Let it be stressed that Pathfinder and Sojourner could not have detected life even if it had existed; they were not programmed to do so, but what they found out would be invaluable when it became possible to send a sample-and-return probe to collect material from the Martian surface. It was all part of a very carefully-planned campaign.

Both Pathfinder and Sojourner operated for much longer than had been expected, but they could not continue indefinitely. The last really good transmission was received on 27 September 1997, but by then it was clear that the power was failing. There were brief, fragmentary signals on 2 October, and again on 8 October, but then Pathfinder 'went silent' – and of course Sojourner no longer had a relay to send its messages back to Earth. For some time Mission Control went on trying to regain contact, but finally, on 4 November, everyone had to admit that the programme had come to an end. Altogether the lander had sent back 16,000 images, and 550 had been received from Sojourner, plus the analyses of the rocks and a vast amount of miscellaneous information. It had been a truly impressive performance.

Much had been learned. For example, it was found that the morning 'fogs' were not fogs at all, but were due to clouds; the temperatures were about ten degrees higher that those measured by the Vikings, and the question of whether or not the Ares Vallis area was an old flood-plain had been finally settled. But above all, perhaps, it had been shown that the 'bounce' landing technique really did work. The comments made by Dr. David Baltimore, President of the California Institute of Technology, summed up the whole situation: 'This mission has advanced our knowledge of Mars tremendously, and will surely be a beacon of success for upcoming missions to the Red Planet. Done quickly and within a very limited budget, Pathfinder sets a standard for twenty-first century space exploration.'

Throughout the first part of 1997 the Mars Global Surveyor (MGS) probe was speeding across space toward the planet, and it entered Mars orbit on

11 September, just as Pathfinder was nearing the end of its glorious career. MGS was a very different sort of vehicle. It was purely an orbiter, with a body approximately five feet square and nearly ten feet long, with solar panels which would, when extended, give a span of 40 feet from tip to tip, and would provide a total power output of 980 watts. MGS was a three-axis controlled vehicle, and was scheduled to last for a long time; the main mapping operation was to start in March 1998 and go on until at least January 2000, though in fact the mapping was delayed for a year because of technical problems (one of the solar panels did not deploy properly). The aim was to explore Mars in all its aspects, including the atmosphere, surface features, mineral distribution and magnetic properties. Up to that time it had been generally believed that Mars had no magnetic field at all.

One very important problem concerned the absence of surface water. Mars had once been wet – Pathfinder had confirmed that – but where had all the water gone? Of course there was plenty of ice, locked up in the polar caps and probably also as permafrost below the surface, but there was still a vast quantity of water unaccounted for. Either it had escaped, or else it had 'gone underground', and although the presence of a subterranean sea did not seem very likely it was possible that there might be more underground ice than had been predicted. It was hoped that MGS would help in clearing up these riddles. First, however, it had to be put into the right orbit.

Because the Delta-2 launcher had limited power, MGS was to be put into its final path by using the method of air-braking. The initial orbit round Mars was highly elliptical, but at each closest approach the space-craft dipped into the upper atmosphere and was slowed by friction, so that successive closest approaches would be lower and lower, and this would continue until eventually MGS was moving round Mars in a circular path around 250 miles above ground level. The problem with the solar panel meant that this could not be achieved until March 1999, a year behind schedule, but the overall programme was not badly affected.

There were several experiments on board, each designed for a specific investigation. The Mars Orbital Camera (MOC) was to produce a daily wide-angle image of Mars, very like the weather photographs of the Earth shown on television news bulletins, while the narrow-angle lens would be capable of imaging objects no more than seven feet long. The Mars Orbital Laser Altimeter (MOLA) would measure heights and depressions; a laser would fire pulses downward ten times per second, and the time taken for the signal to go to the surface and back would make it possible to produce elevation maps precise to within a few tens of feet. The Thermal Emission Spectrometer (TES) would analyze infra-red radiation from the surface, and measure temperatures as well as giving information about the composition of the rocks and dunes. The Magnetometer Electron Reflectrometer (MER) would search for any magnetic field. The complement of instruments was completed by the Mars Relay and the radio science equipment.

The first major discovery was made on 15 September, only four days after orbit insertion: MGS detected a magnetic field. It was weak, with a strength of no more than 1/800 of the Earth's field, but it did seem to be there; earlier indications of it from Russian probes had been inconclusive. Later it was found that the magnetic situation is somewhat complex, and there are localized 'magnetic anomalies' in the crust. The overall field has an orientation similar to that of the Earth, so that in theory a magnetic compass would always point north, but the anomalies could often override this, so that navigating by an ordinary compass would be a very hit-or-miss affair. Whether an iron-rich core still acts in the fashion of a dynamo, as happens on Earth, is still not clear.

The images sent back from MGS were superb from the outset, and showed details which had been too delicate to be previously observed. For example, there were views of the canyons in the Noctis Labyrinthus area which showed evidence of layering, while flooded craters were also seen, and it was evident that many of the surface features had been shaped by winds. The polar caps were found to be thick, with depths of several thousands of feet – a far cry from the time, not so long ago, when they had been dismissed as mere coatings of hoar-frost. Large areas of the caps were surprisingly smooth; the MOLA equipment could measure altitudes to an accuracy of only a foot or two. And finally, MGS disposed of the Martian 'face' in Cydonia. Seen from a different angle, and in more detail, all semblance of a human form disappeared.

MGS is still operating perfectly as I write these words (August 1998). The best is yet to come.

Up to very recently only the Russians and the Americans had launched probes to Mars, but on 3 July 1998 Japan joined in, sending up a lightweight space-craft known as Planet B. It was 4 feet 10 inches wide and 1 foot 9 inches high, and was designed purely as an orbiter. It was launched by a powerful M5 rocket, and is scheduled to remain in orbit round the Earth for several months before being sent out to Mars, where it will stay in a closed path round the Red Planet for two years. Its aim is to make studies of the Martian surface as well as the space environment.

It is too early to say whether or not the attempt will be successful, but so far Japan's space record has been good, and Planet B (now re-named Nozomi, which is Japanese for 'Hope') may well make a valuable contribution. Nowadays, the exploration of Mars has become truly international.

In admiring the triumphs of Pathfinder, Sojourner and MGS, it is unfair to forget the work of a different kind of space-craft: the HST or Hubble Space Telescope, circling the Earth and sending back a steady stream of information. Of course it is concerned mainly with 'deep space', but excellent pictures of Mars have also been taken, far surpassing anything which can be obtained from ground level. Moreover, HST will certainly continue to go on working for much longer than MGS, so that it will be able to keep Mars under regular surveillance. Yet all in all, we have to admit that the future of Martian exploration lies with the rocket. We still have much to learn before we are ready for the first manned mission, but at least the space-craft sent there in the last years of the twentieth century have shown us the way.

CHAPTER 14

Phobos and Deimos

Most people, I suppose, have read Jonathan Swift's classic *Gulliver's Voyages*. To be more precise, most people have read the voyages to Lilliput (the country of the midgets) and Brobdignag (the country of the giants). The remaining Voyages are less famous, but one of them is particularly relevant in a discussion of Mars. Dr. Lemuel Gulliver is said to have visited Laputa, an airborne island which probably qualifies as the first fictional flying saucer. The Laputan astronomers were so skilful that they had discovered 'two lesser stars, or satellites, which revolve about Mars, whereof the innermost is distant from the centre of the primary planet exactly three of his diameters, the outermost five; the former revolves in the space of ten hours, the latter in twenty-one and a half'. In other words, the inner satellite moves round Mars so quickly that it completes its circuit in less than a sol.

The Voyage to Laputa was written in 1727. At that time there was no telescope in existence powerful enough to show the two satellites of Mars that we now know to be genuine, and this was also true when Swift died in 1745. Five years later, the French novelist Voltaire wrote a rather strange story, *Micromégas*, in which the Solar System is visited by a being from the star Sirius. Voltaire also credited Mars with two moons.

In fact, the Martian satellites were not discovered until 1877, and the Swift and Voltaire stories have led to some peculiar speculations. It has even been suggested that our remote ancestors had optical instruments which enabled them to track the satellites down. Unfortunately for this intriguing idea, the true explanation is very simple, and Voltaire himself wrote it down. He pointed out that because Mars is further away from the Sun than we are, how can it possibly manage with less than two moons?

There was also something of a progression in the numbers of planetary satellites known in the mid-eighteenth century. Mercury and Venus appeared to be unattended. (They still are. A satellite of Venus has been

reported now and then, but is nothing more than a telescopic 'ghost'.)
The Earth, of course, had one moon. Jupiter had four known satellites,
all discovered by the earliest telescopic observers; Galileo saw them in
1610, and others detected them at about the same time. Saturn had a
retinue of five: Titan, discovered by Christiaan Huygens in 1655, and
Iapetus, Rhea, Dione and Tethys, all found by Giovanni Cassini between
1671 and 1684. So there was the progression: Venus 0, Earth 1, Jupiter 4,
Saturn 5. It was logical to give two attendants to Mars, and this is what
both Swift and Voltaire proceeded to do.

As larger telescopes were built, more planetary satellites came to light.
By 1850 the grand total was eighteen – Earth 1, Jupiter 4, Saturn 8, Uranus
4 and Neptune 1. Yet only the Earth among the inner group of planets seemed
to be accompanied. In 1783 William Herschel made an unsuccessful search
for a Martian satellite, and in 1862 and 1864 Heinrich d'Arrest, at the Copen-
hagen Observatory, was equally luckless. The general feeling among astronomers was that the poet Tennyson was right in describing 'the snowy
poles of moonless Mars'.

Then came the close opposition of 1877, when Schiaparelli drew his map
of Mars and described the canal network for the first time. Over in the United
States, a well-known observer named Asaph Hall decided to renew the
attempt to find a satellite. He was well equipped for the search, since he was
able to use the 26-inch refractor at Washington – one of the largest telescopes in the world at that time, and also one of the best (the object-glass
was made by Clark, whose skill was second to none). Hall began work in
early August. At first he was as unsuccessful as Herschel and d'Arrest had
been, and apparently he was on the verge of giving up when his wife persuaded him to continue for at least another few nights.

On 10 August he began observing as usual. For some time he saw nothing
unusual about the background of stars, but at 2.30 in the morning of 11
August he caught sight of a very faint object close to Mars which seemed to
be much more promising.

Unfortunately, fog rising from the nearby Potomac River came up before
he had had time to do more than make a quick observation of the suspected
object, and the next four nights were useless, as cloud and mist prevailed.
Finally, on 16 August, the weather cleared. Hall was able to recover his
suspected satellite, and he saw that it was moving along with Mars, so that
it was a true attendant. On the following night there were startling develop-

ments. The original satellite was seen again, and another was found, even closer-in to Mars.

Hall's announcement, made on 18 August, caused a great deal of interest, which increased when it became clear that the inner satellite at least was a most remarkable object. In Hall's own words, written a few days later: 'At first I thought that there were two or three moons, since it seemed to me at that time very improbable that a satellite should revolve around its primary in less time than that in which the primary rotates. To decide this point, I watched the moon throughout the nights of 20 and 21 August, and saw that there was in fact but one inner moon, which made its revolution around the primary in less than one-third the time of the primary's rotation, a case unique in the Solar System.'

Certainly this was very strange. The inner moon – actually the second in order of discovery – proved to have a revolution period of a mere 7 hours 39 minutes. Therefore, reckoning by this satellite, the Martian month was shorter than the sol. Swift had been right, even though his description had been no more than a shot in the dark.

Hall, as discoverer, had the honour of naming the satellites. He wrote: 'Of the various names that have been proposed . . . I have chosen those suggested by Mr. Madan of Eton, England, viz. Deimos for the outer satellite: Phobos for the inner satellite. These are generally the names of the horses that drew the chariot of Mars; but in the lines referred to (in the fifteenth Book of Homer's *Iliad*) they are personified by Homer, and mean the attendants, or sons of Mars . . . "He (Mars) spake, and summoned Fear (Phobos) and Flight (Deimos) to yoke his steeds".'

Phobos and Deimos they have remained. Almost a century later, Mariner 9 photographed craters on each. The main carter on Phobos has been named Stickney, because the maiden name of Hall's wife was Angeline Stickney. And the two most prominent craters on Deimos have been named, appropriately, Swift and Voltaire.

Encouraged by Hall's success, other astronomers set to work to search for new satellites. They failed. In much more recent times searches were carried out by G. P. Kuiper, using the 82-inch reflector at the McDonald Observatory in Texas during the oppositions of 1952, 1954 and 1956. He concluded that no new satellite as much as a mile across could exist, and we can now be sure that Phobos and Deimos are the only natural attendants.

Both are tiny, which explains why they were not discovered until 1877, and both are irregular in shape; Phobos measures 16.8 × 13.7 × 11.2 miles, Deimos only 9.3 × 7.5 × 6.2 miles. From Earth they look like tiny specks of light, and before the voyage of Mariner 9 we knew nothing about their surface features. Their stellar magnitudes are 12.1 for Phobos and 13.3 for Deimos. This is not particularly faint, but the satellites are so close to Mars that they are drowned in the glare; Phobos can never be more than 20 seconds of arc away from the Martian limb, and Deimos 65 seconds, which is not very much. With a moderate telescope, the only way to see them is to block out the disk of Mars with a device known as an occulting bar. I have managed to see both satellites with my 15-inch reflector, but only under exceptionally good conditions, and even then with some difficulty. Antoniadi claimed that Phobos was white and Deimos bluish, but I fear that he was drawing upon his imagination. To see a colour in a spot of light as dim as Deimos is, frankly, impossible.

It is plain that Phobos and Deimos are quite unlike our own massive Moon. Their escape velocities are low – around 30 m.p.h. for Phobos and 15 m.p.h. for Deimos – so that bringing in a space-ship would be more in the nature of a docking operation than a conventional landing, and an astronaut standing on the surface of either satellite would have practically no feeling of weight.

Phobos moves in an almost circular orbit, at a distance of 5800 miles from the centre of Mars, and in the plane of the planet's equator. This means that it is no more than 3700 miles above the Martian surface, which is roughly the same distance as that between London and Aden. An observer standing at a high latitude on Mars would never be able to see Phobos at all, since it would remain below the horizon from latitudes greater than 69 degrees north or south. (Note, though, that it can be seen from both the Viking sites and from that of Pathfinder, since Chryse, Utopia and Ares Vallis are all closer to the equator than this limit.)

Antoniadi worked out what our hypothetical Martian observer would see, and he was to all intents and purposes right. From the equator of Mars, Phobos would rise in the west, cross the sky in only 4¼ hours, and set in the east, during which time it would go through more than half its cycle of phases from new to full. The interval between successive risings would be 11 hours 6 minutes, and Phobos would seem appreciably larger when high up than when low down. Often it would be eclipsed by the shadow of Mars – in fact, about 1330 times every Martian year, and it would be shadow-free

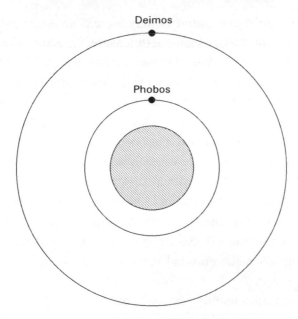

Orbits of Phobos and Deimos, to scale.

for its whole passage across the sky only for restricted periods near mid-summer and midwinter. Even then it would be fairly useless as a source of illumination at night.

Deimos, moving at a distance of about 12,500 miles above the surface of Mars – that is to say, about the distance from Australia to England – has a revolution period of 30 hours 18 minutes. It, too, has an orbit which is virtually circular and in the equatorial plane. As Mars spins, Deimos almost keeps pace with it, so that it remains above the horizon of the Martian equator for sixty hours consecutively, during which it passes through its phase-cycle twice. It, too, is invisible from the polar regions; any observer above latitude 82 degrees north or south will never see it. It would be eclipsed 130 times every year, but in any case its phases would be none too easy to see with the naked eye. It would look rather like Venus as seen from Earth, though rather larger and appreciably dimmer.

The two moonlets undergo all sorts of eclipse and occultation phenomena, rather as though playing cosmic hide-and-seek. They can, of course, pass in front of the Sun, but they are too small to blot out the disk, so that solar eclipses can never occur. Phobos can hide one-third of the Sun, and passes across the disk in 19 seconds; this happens 1300 times a year. Deimos makes 120 crossings, but covers only one-ninth of the disk, and takes two minutes to pass right over. Future observers will no doubt enjoy watching the spectacle of both satellites silhouetted against the Sun at the same time, as can happen quite often. Also, Deimos can be eclipsed by Phobos.

I cannot resist mentioning a peculiar theory proposed in 1959 by Iosif Shklovsky, who was undoubtedly one of the world's great astronomers, but whose views about extraterrestrial life were, perhaps, a little extreme. There had been suggestions that Phobos was speeding up in its path, so that it was gradually spiralling downward and might be expected to crash-land on Mars in from 35 to 40 million years from now. The calculations were made origi- nally by B. P. Sharpless in 1945, and had been supported by various other astronomers. It was also noted that although Phobos was outside the Roche limit for Mars – that is to say the 'danger zone, inside which a fragile body would be disrupted by the gravitational pull of the planet – it was not very far outside.

Shklovsky caused something of a sensation by his claim that this speed- ing-up was due to friction against the Martian atmosphere. He went on to point out that at a height of more than 3000 miles above the surface, this atmosphere must be very thin indeed, so that it would be unable to influence a moving satellite of appreciable mass. Therefore, reasoned Shklovsky, the mass of Phobos must be negligible. It follows that it must be hollow, and so presumably of artificial construction – in fact a space-station, launched by the Martians for reasons of their own! The idea was eagerly seized upon by flying saucer enthusiasts, and it was even suggested that Herschel and d'Arrest had failed to find Phobos (or Deimos) for the excellent reason that they had not been sent up from Mars much before 1877.

Astronomers in general were not impressed, and members of the Soviet Academy of Sciences were frankly embarrassed; in Moscow, efforts were made to play down the whole episode. Years later, when space research methods had shown without doubt that Phobos and Deimos were natural bodies, I asked Shklovsky about it. He told me that he had never meant to be taken seriously, and that his paper had been nothing more than a practical joke. I think it might be wisest for me to say 'No comment . . .'.

Until 1969 our ignorance about the physical characteristics of Phobos and Deimos was complete. The first direct information came from Mariner 7. A photograph taken during the fly-by showed Phobos silhouetted against the disk of Mars, and it was seen that the form was irregular. Mariner 9 took pictures from close range, and found that both satellites are roughly triaxial;

they give the impression of being fragments of rocky material which could well have been broken away from a larger body.

We now know that both Phobos and Deimos have what is termed captured or synchronous rotation; that is to say, their orbital periods are the same as their rotation periods – approximately 7½ hours for Phobos, 30½ hours for Deimos. This means that each satellite will keep the same face turned toward Mars all the time. An observer on Mars will never be able to see the whole of the 'far side' of either satellite.

There is nothing surprising in this, and our Moon behaves in the same way with respect to the Earth. Until 1959, when the Russians sent their camera-carrying probe Luna 3 right round the lunar globe, we knew nothing positive about the averted hemisphere. Tidal friction is responsible for this state of affairs, and the same applies to the satellites of Mars, which have, so to speak, been 'braked' – though it must be added that the satellites do rotate relative to the Sun, so that all parts of their surfaces are bathed in sunlight at one time or another. Also, the longer axes of the satellites are directed toward Mars. On Phobos, the sub-Mars point – that is to say, the point on the surface of Phobos from which Mars would appear directly over-

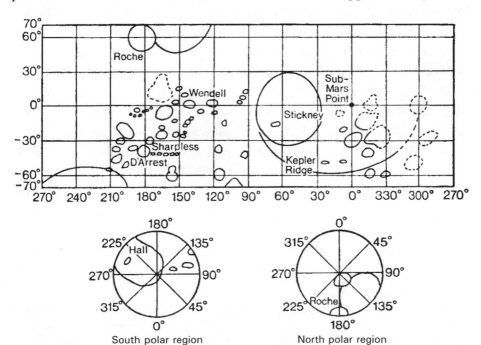

Map of Phobos.

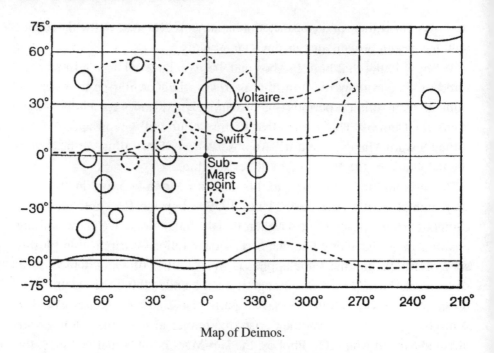

Map of Deimos.

head – is outside the wall of the main crater, Stickney; on Deimos the sub-Mars point is not very far from Swift and Voltaire. To an observer on Phobos, Mars would subtend an angle of 42 degrees in the sky, so that it would indeed be impressive; even from Deimos, so much further out, Mars would still be dominant.*

Both Phobos and Deimos were imaged from the Viking orbiters, and indeed the only close-range pictures taken since then were those of the Russian probe Phobos-2 in 1989. Both satellites are cratered, but they are not alike. Phobos shows strange grooves, inclined to the satellite's equator by about 30 degrees; they are around 300 to 400 feet wide, and in some cases over 60 feet deep. They seem to be associated with the formation of the largest crater, Stickney, which is five miles across. Since the diameter of Phobos is less than seventeen miles, it seems that if Stickney were formed by meteoritic impact – as is generally believed – Phobos must have been in grave danger of being shattered, particularly as it is of low density; its mate-

---

* When discovered, Phobos was unique in having an orbital period less than the rotation period of its primary planet. This is no longer so; the same is true of the two innermost satellites of Jupiter, Metis and Adrastea. Jupiter's axial rotation period is just less than ten hours; the orbital periods of Metis and Adrastea are little over seven hours. Adrastea is about the same size as Phobos, Metis somewhat larger. Both seem to be asteroidal in nature.

rial may be similar to that of meteorites known as carbonaceous chondrites, and is not more than twice as dense as water. There are also strings of tiny pits which look remarkably like blowholes. It is covered with a loose layer or regolith, a few inches thick at most. Incidentally, it does seem to be true that Phobos really is spiralling downward, so that it will eventually crash on to the Martian surface. It will make a huge crater, but the impact is not now thought to be due for around 50 million years, so that future colonists on the planet will have ample time to prepare for it.

Deimos, much smaller than Phobos, has a deeper regolith, so that its surface features are more subdued, and all the craters are very small indeed. Voltaire is much the most prominent of them.

Midgets though they are, the Martian satellites are interesting objects, and in the 1980s the Russians decided to send two probes to Phobos. Both were elaborate. The Phobos-1 space-craft carried a total of 31 experiments, designed to study not only Phobos but also Mars itself. It was relatively massive – its initial weight was 2600 kilograms – and it was powered by solar panels in the usual way. After a 200-day cruise it was scheduled to go into orbit round Mars, and following a series of manœuvres to end up in a path about 200 miles above that of Phobos. It would approach Phobos to within a range of two miles, and would carry out close-range surveys; it would also drop two small landers on to the surface, fastening them down by what might well be termed cosmic harpoons. These landers would, it was hoped, operate for at least two months.

Phobos-1 was launched from the Baikonur cosmodrome on 7 July 1988. At first all went well, but on 2 September contact was lost permanently. There was no mystery about this; it was due to human error. A faulty command was sent out from the U.S.S.R. Mission Control, and equipment on the space-probe failed to override it. In effect, Phobos-1 was told to switch itself off, which it duly did. The technician responsible must have become rather unpopular; significantly, he was never named.

Phobos-2, launched from Baikonur by a Proton rocket on 12 July, was even more ambitious. As well as duplicating most of the experiments of its predecessor, it was intended to drop two small vehicles. One would simply anchor itself down, but the other would be spring-loaded, so that it could literally hop around on Phobos' surface.

Again all went well at the outset, and as Phobos-2 approached its target some reasonably good images were obtained, both of Phobos itself and of

Mars. Yet again the Fates were unkind. On 27 March 1989, well before the most important part of the mission could begin, contact was lost. This time the reason was assumed to be a fault in the on-board computer.

Every effort was made to locate the errant space-craft, and searches were carried out not only in the U.S.S.R. but also elsewhere. At La Silla Observatory, in the Atacama Desert of Northern Chile, an optical search was made with the powerful Danish telescope, but there was no luck; Phobos-2 had vanished, and nothing more was ever heard from it. It cannot be written off as a total failure, because it did produce some fairly good images and also gave the first vague indications of a Martian magnetic field, but certainly its main task was unfulfilled.

New temperature data and close-range images of Phobos obtained in September 1998 by Global Surveyor show that the surface must be composed largely of finely ground powder at least three feet thick. Measurements of the temperatures on the day and night sides show that the maximum temperature on the sunlit side is 25 degrees Fahrenheit (−4 degrees Centigrade), while on the dark side the temperature is −170 degrees Fahrenheit (−112 degrees Centigrade). The small particles lose their heat very quickly when plunged into darkness. According to Philip Christensen, one of the main investigators, 'this has to be an incredibly fine powder formed from meteoroid impacts over millions of years, and it looks as if the whole surface is made up of fine dust'. Global Surveyor also took new pictures of Stickney, Phobos' largest crater, which showed light and dark streaks trailing down the slopes of the bowl, illustrating that even in a gravity field no more than 1/1000th of that of the Earth, débris still tumbles downhill. Large boulders appear to be partly buried in the surface material.

There is absolutely no similarity between the Martian satellites and our own massive Moon, and it is fairly certain that Phobos and Deimos are not genuine satellites at all; much more probably they are ex-asteroids, which wandered so close to Mars that they were unable to break free. This sounds reasonable enough in view of the fact that Mars is much closer to the main asteroid belt than we are. Many small asteroids – those of the Amor and Apollo groups – cross the orbit of Mars, and could well have been captured long ago, though it is true that this would require very special circumstances. As we have noted, there is also one asteroid – Eureka – which moves in the same orbit as Mars, though it keeps well clear of the planet and is in no danger of collision.

One day, no doubt, Phobos and Deimos will be pressed into service as natural space-stations, and certainly they will be very convenient. They will then be of real importance to us, even though we have regretfully had to abandon the idea that they could be made up of metal rather than conventional rock.

CHAPTER 15

The Martian Base

Apart from the Moon, only two worlds are within the range of present-day technology insofar as manned space-travel is concerned. These are Venus and Mars. All the rest are either too far away or else too hostile. Titan, the largest satellite of Saturn, does have a dense atmosphere made up chiefly of nitrogen, but is so remote that any talk of sending an expedition there is, to put it mildly, premature, and in any case the atmosphere contains so much methane that we would hardly find it welcoming.

Venus, with its intense heat, choking atmosphere and clouds of sulphuric acid, may be ruled out. So, as always, we come back to Mars, and plans for an expedition there are already being made. It will be surprising if a fully-fledged Martian base is not set up within the next fifty years.

When will the first party set out? At a recent General Assembly of the International Astronomical Union, the controlling body of world astronomy, the favoured date was around 2020, and there is nothing far-fetched about this. Travel to Mars now seems more plausible than reaching the Moon did when I first became interested in astronomy in the late 1920s.

Yet there are problems which could change the whole picture. Probably the main one concerns the human body. Men have remained in space, on the Russian space-station Mir, for over a year and have come down safely, but it is too early to be sure about any long-term effects. Moreover, on Mir a cosmonaut with physical problems could hastily be brought home (as happened on more than one occasion), but this would not be possible on a voyage to Mars. Of course, all precautions will be taken during the journey – regular exercise, for example – but it would be foolish to claim that all these precautions will be fully effective, and the only way to find out for certain is to try. There is also the question of radiation. Solar storms release large amounts of short-wave radiation and charged particles which are highly dangerous, and whether a space-craft can provide really effective screening is debatable. Remember, a journey to Mars is bound to take months, whereas an Apollo expedition to the Moon can be completed is less than a fortnight.

There is another vital difference between travelling to the Moon and travelling to Mars. A lunar mission involved one space-craft, crewed by three men. A Martian expedition will need several space-craft and many more crew members, so that the new 'Martians' – women as well as men – will have to be very carefully selected, and there will also have to be the equivalent of a space hospital, as well as trained medical staff. Space is always a dangerous environment; accidents can – and do – happen, so that on a prolonged journey it will be essential to have specialists capable of dealing with any emergency.

At least there will be no difficulty in keeping contact with Earth, and by the time that the first expedition leaves we will know definitely whether there is or ever has been any trace of life on Mars. Neither will there be any water shortage; there is plenty of ice, and in this respect Mars is far more co-operative than the Moon.

These facts are obvious enough, but when we come to consider the expedition itself, and the way in which a Base may be established, everyone is bound to have his own ideas. The comments which follow are my own; I do not pretend to be an expert in these fields, and I may prove to be very wide of the mark, but all I can do is to present my own views.

First, the journey. Where will it start? Quite possibly from the Moon; by the time we are ready to set off for Mars, there may well be a proper station on the lunar surface, and of course this will make the departure much easier. As for the cruise – well, it will take a matter of many weeks, and the crew members will have to take care to keep themselves in peak condition. Once on Mars there can be no chance of a quick 'turn-around'; it will be necessary to wait until the two planets are suitably placed for the return journey.

It is quite likely that the first step will be into a closed path round Mars, and here the satellites may come in useful, particularly Deimos, which is so small and lightweight that its gravitational pull is negligible; 'landing' on it would be more in the nature of a docking operation. A supply depôt might be left there, together with radio equipment, so that Deimos could serve not only as a communications relay but also as a beacon for navigators on the Martian surface. Mars would be a superb sight in the sky, over thirty times as large as the Moon appears from Earth; the brilliant Mars night would cast a strange, ruddy radiance over the miniature world, and the famous dark areas on Mars itself would be well displayed. They would seem to shift only slowly, because Deimos almost keeps pace with Mars as the planet rotates on its axis. Phobos, much close-in, would scurry across the sky of Deimos, and might be rather less valuable.

Then would come the landing itself. By the time of the expedition, most of the present-day problems will have been solved, and there is no reason to doubt that touch-down will be more or less routine (if anything in space can be classed as 'routine'). Very probably a major cargo vessel will be one of the Mars fleet, since most essential materials will have to be brought from Earth. But once on Mars, the sensation of weight will return, and after so long a period under conditions of zero gravity this will certainly be a great relief.

If the atmosphere of Mars were dust-free, the sky would presumably be very dark blue, but – as the twentieth century probes have told us – the actual colour of the daytime sky is pink. The nights will be very dark, and neither Phobos nor Deimos can provide much in the way of illumination; the stars will be brilliant, and will twinkle much less than they do from Earth. The constellation patterns will be exactly the same, but the positions of the celestial poles will be different; the north pole is not far from Deneb in Cygnus, while the south pole will be found among the stars in the pattern known to us as the False Cross. Earth will be an inferior planet showing phases, as Venus appears to us; Venus will be reasonably prominent when at its best, but Mercury will remain so close to the Sun that it will be very hard to glimpse with the naked eye. Jupiter and Saturn will be bright, and Uranus will be well above naked-eye visibility; so too will a number of asteroids.

As for the design of the pioneer Base – well, all sorts of suggestions have been made, and it is still early to decide upon the final choice. It has been claimed that radiation problems will mean living underground rather than on the surface, but this does not seem to be essential. We are not sure whether the atmosphere will provide an effective shield against harmful cosmic and short-wave radiation, and there is no equivalent of our ozone layer, but Mars is fifty million miles further from the Sun, and it may be that the dangers have been exaggerated. Remember the alarms of the early 1950s, when it was claimed that any astronaut venturing above the top of our protective atmosphere would at once be seared by cosmic rays and left gasping as his craft disintegrated under a hail of meteoritic bullets. Not until Yuri Gagarin's epic flight in 1961 were these 'bogeys' finally laid to rest, and it may be that there is a lesson to be learned here.

As we have noted, the Martian atmosphere, tenuous though it is, is quite sufficient to burn up meteors and produce shooting-star effects. It is not sufficient to give protection against larger bodies, but neither is ours; a large impactor will crash through the Earth's air and land forcefully enough to make a crater such as that in Arizona. There is always the chance of a major strike, and the same is true of Mars, but this is something about which we can do nothing at all, and the risks are not really very high.

The classic picture of a Martian Base is of a colony made up of graceful domes rising from the desert landscape, kept inflated by the pressure of air inside them. There is nothing outrageous in this, so let us assume that the first Base is indeed planned along these lines. The material of the dome must be able to withstand not only stress, but also the wide variations in temperature. There will be no problems with intense heat, as there would be on Venus; even on the Martian equator the temperature never matches that of a June day on London, but the atmosphere is very poor at shutting in warmth, and the nights are intensely cold.

Obviously the choice of site will be important, and here we must await the results of further research; for example, some of the old flood-plains may have useful ice not far below the surface. I will not be surprised if the first Base is erected close to the site of the Sagan Memorial Station, which could well become the official Martian headquarters.

The worst danger to the colonists will always be that of an air leak. The planet's atmosphere is so thin that quite apart from the fact that we could not breathe it, the pressure is so low that an astronaut not wearing a proper suit would promptly burst as the blood inside him boiled – literally as well as metaphorically. Older science-fiction novels showed men walking around at Martian noon with no protection other than an oxygen mask, but these stories were written before we realized that the air is as tenuous as it actually is.

Certainly there will be several units in the pioneer Base, and each unit will have to have its own system of airlocks for entry and exit; no doubt there will be connecting passages which can be sealed off quickly if need be. There is always the chance of a meteorite strike powerful enough to puncture the outer skin, and there must be somewhere for the inhabitants to go until repairs can be carried out.

What about dust-storms? These occur quite often, and there are also dust-devils; there may even be Martian equivalents of our tornadoes. But here we come back to the thinness of the atmosphere, and even the strongest Martian 'twister' is not likely to be violent enough to cause real damage to a well-planned and well-built Base. Ground tremors – Marsquakes – are likely to be so feeble that if the equivalents hit Japan, for instance, they would not be regarded as worth recording at all.

The first Base must be purely scientific, and there will not be much time for relaxation, though with under one-third Earth gravity it will be essential to take proper exercise, and there will undoubtedly be what we can only call a Martian gymnasium. Reports will be sent back to Earth daily, at least, though it is true that when Earth and Mars are on opposite sides of the Sun (that is to say, when Mars is at superior conjunction as seen from Earth) things will be less straightforward, and the messages will have to be routed by some relay station in solar orbit. Only when the Base has been in operation for some time will it be possible to decide how things will develop. There is always a chance – though a slim one – that some unexpected hazard will make the whole project untenable, so that Mars will be abandoned, but it is much more likely that other expeditions will arrive, so that before long there will be one or more flourishing colonies. Travel between different parts of Mars should present no real difficulty, though the tenuous air means that rocket planes will be more in vogue than aircraft of the kind we use at home.

Let us assume that the Base has been established, has been successful and has been extended. What will life there be like, and what will be the overall benefits to mankind?

There is no reason why life inside the Base should not be fairly normal. One-third gravity will be comfortable enough; the astronauts had no problems on the Moon, where the gravity is much weaker than on Mars. Things will not float around, as they do under conditions of weightlessness, and, for example, there will be no need to fasten yourself to your bed before going to sleep in order to guard against drifting away. There will be no sanitation problems, as there are inside space-stations.* Plants will be able to grow inside an airtight dome, and will provide food; it is not likely that Earth-type plants can be persuaded to take root in the open, but there might be a chance

---

* One of the most frequently-asked questions, particularly by young enthusiasts, is: 'How does one go to the bathroom in space?'

of what is known as hydroponic farming, in which plants are suspended in a net above channels containing nutrients. Whether anything of the sort will in fact be practicable remains to be seen.

This leads on to another question of paramount importance: can there be anything harmful on Mars? When the first astronauts came back from the Moon they were strictly quarantined until it was certain that they had brought back no dangerous contaminants, and only after the second mission was quarantining abandoned, because it was so patently obvious that the Moon is, and always has been, sterile. Quarantining will be applied even more strictly to the first samples brought back to Earth from Mars, as ought to happen within the next few years. Mars, unlike the Moon, has an atmosphere, and we cannot rule out the chance that there may be primitive life-forms. I would say that the dangers of any harmful Martian organisms are infinitesimal, but I agree that we must make absolutely certain before we bring any samples down to the Earth's surface. It is also possible that spacecraft from Earth which have landed on Mars have already caused contamination there, even though before launch they were sterilized as effectively as possible. But again I would say that the chances are more than a million to one against.

As time goes by, and the Bases expand, other factors will have to be taken into account. Recreation – essential. Most sports will be possible, though with modifications; as a cricketer I have a vivid picture of a match being played under one-third gravity, and speaking as a leg-break bowler I have the feeling that it will work very much to the advantage of the batsman, who will be able to hit the ball further and far higher than even W.G. could have done at Lord's. Tennis, football and the rest will no doubt be played, but one of the main Martian sports of the future must surely be mountaineering. There is plenty of scope here, and the handicap of having to wear a pressure suit will be offset by the reduction in Earth weight; neither will a fall be so dangerous. Who will be the first Hillary or Tenzing to climb to the top of Olympus Mons, or the peaks of Tharsis?

There are still people who ask why it is logical to spend money on space research, when there is so much to be done at home. Yet in fact the benefits will be great in all branches of knowledge, including medical research. It is also true that any project such as a Martian Base must be international, so that in going to Mars we will be taking a major step toward uniting the Earth. I am reminded of a remark made by a famous last-century scientist when asked what was the use of the new-fangled science of electricity. He replied: 'Madam, what is the good of a new-born baby?'

Tourism to Mars may come eventually, though certainly not until well into the twenty-first century. There will also be children born and brought up on Mars, and this brings me to another point which I have seldom seen stressed elsewhere. Will a boy or girl born and reared on Mars ever be able to come to Earth, where the pull of gravity is so much greater? Imagine how you would feel if, without any increase in your muscle power or your heart strength, your weight suddenly increased by a factor of three. You might not be able to cope. I am prepared to believe that within the next hundred years there may be two entirely separate branches of Homo sapiens: Earth-dwellers, and our kinsfolk from Mars who can never come to Earth at all, though naturally they would be quite happy on the Moon. This may sound like science fiction in 1998, but by 2098 it may have become science fact.

Perhaps this is about as far as we can go at the present time. Innumerable difficulties remain to be overcome, but unless there are any major hazards, either natural or of our own making, the new century really will see us established on Mars. The Red Planet presents us with a challenge; it is a challenge which mankind will not ignore.

Above: Sand dunes in Chryse; Viking 1 lander.

Below: Trench dug by Viking 1. Material was drawn back into the space-craft and analyzed, but no definite signs of life were found.

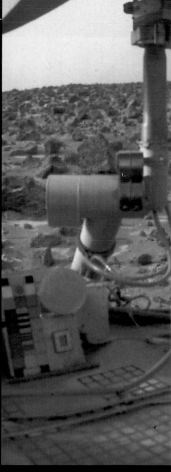

Above: Sunset over the Viking 1 landing site. The temperature is falling very quickly.

Right: Frost at the site of the Viking 2 lander, in Utopia. Part of the lander itself is shown. The area lay well south of Chryse, where VL1 had come down.

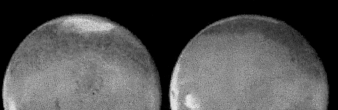

Left: Mars in 1997, from the Hubble Space Telescope. South is at the top. The south polar cap and the main dark areas are well shown, together with some clouds.

Left: Mars Observer, which went out of contact as it approached Mars and was lost.

Above: Meteorite ALH 84001, found in the Antarctic, and believed by many people to have been blasted away from Mars.

Right: Launch of the Pathfinder probe to Mars; it carried the Sojourner rover.

Above: Safe landing; Pathfinder has come to rest on Mars. Sojourner is still inside, waiting to be deployed.

Centre left: Wedge and Flat Top; Wedge is to the left. Directly below Flat Top is a rock nicknamed Bam Bam.

Below left: Panoramic view of the Ares Vallis area; on the horizon are the Twin Peaks, South to the left and North to the right. The rear ramp from which Sojourner has descended is to the left, and above it the first rock to be closely examined, Barnacle Bill.

Left: Sojourner, on Sol 30 (the 30th day on Mars) . . . It is still functional.

Centre left: Barnacle Bill and Yogi, with Sojourner. The two rocks are not identical in composition.

Below: Sunset on Ares Vallis; darkness comes to the old flood-plain, but the sky remains pinkish for some hours because of the atmospheric dust. The Twin Peaks are seen at the horizon.

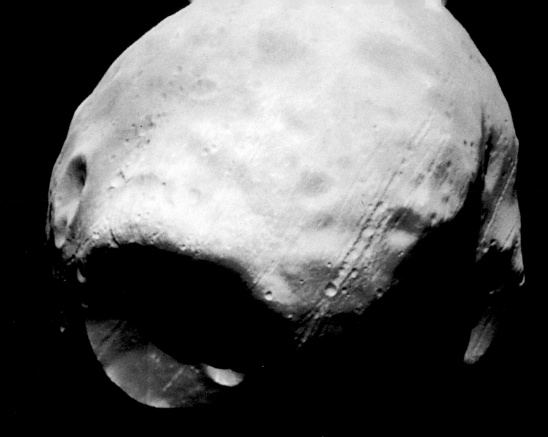

Above: Phobos, from Viking 1. Note the large crater Stickney.

Right: Grooves and craterlets on Phobos; Viking image.

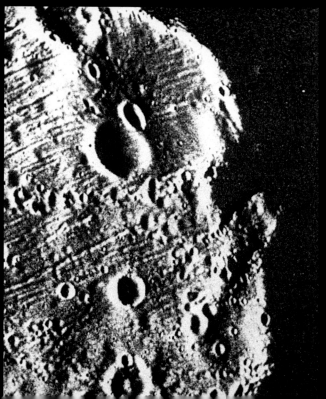

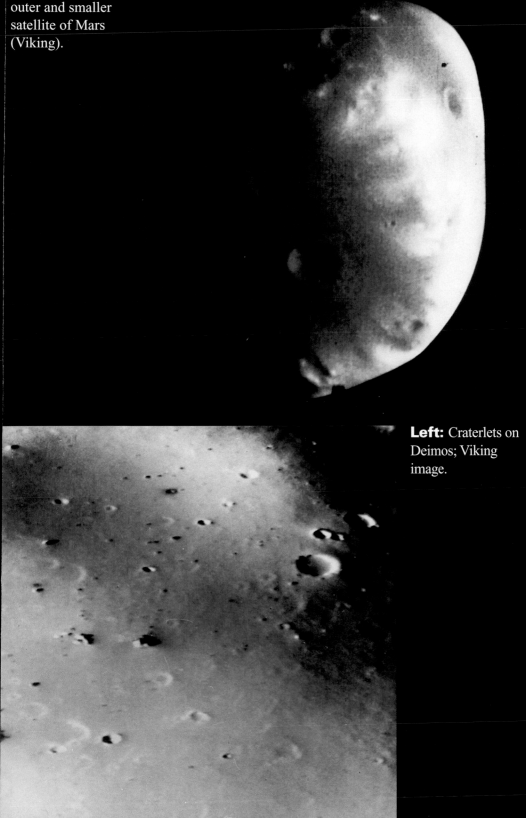

outer and smaller
satellite of Mars
(Viking).

Left: Craterlets on
Deimos; Viking
image.

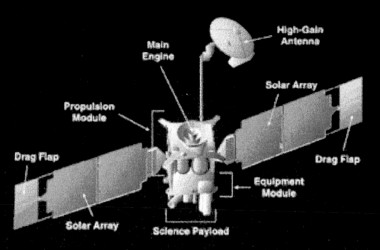

Main
Engine

High-Gain
Antenna

Propulsion
Module

Solar Array

Drag Flap

Drag Flap

Equipment
Module

Solar Array

Science Payload

Left: Mars Global Surveyor, scheduled to carry out detailed mapping of the entire Martian surface.

Right: Noctis Labyrinthus, imaged by Mars Global Surveyor on 19 September 1997, from a range of 370 miles.

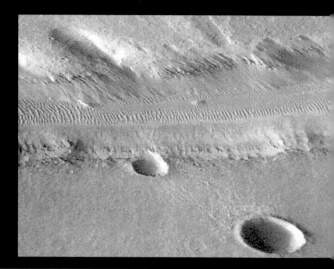

Left: The first Japanese space-probe to Mars, Planet B (now renamed Hope), launched 3 July 1998 and still en route when this book went to press. Illustration by permission of the Japanese Institute of Space and Astronomical Science.

APPENDIX I

Observing Mars

Mars is a difficult telescopic object. This may not seem easy to believe at first sight, because Mars, as we know, can become brighter than any other planet apart from Venus; but it is a fact. Anyone who buys a small telescope, pokes it through a bedroom window and expects to see a network of canals is doomed to disappointment.

Even when it is at its closest to us, the apparent diameter is a mere 25.7 seconds of arc, which is only about half that of Jupiter. And when Mars is at its greatest distance, the apparent diameter shrinks to 3.5 seconds of arc, comparable with that of remote Uranus. Generally speaking, little can be seen on the disk when the diameter drops below 6 or 7 seconds of arc – and at all times a telescope of considerable aperture is needed for any scientific work which may be regarded as useful. Mars is one planet upon which a high magnification has to be used, though this does not mean that one should go too far; a smaller, sharp image is always better than a large but blurred view.

When Mars is reasonably near opposition, a telescope as small as a 3-inch refractor will show the main markings; the polar caps (or, to be more precise, whichever polar cap happens to be turned toward us) and dark areas such as Syrtis Major and Acidalia Planitia. Even here there must be a qualification, because the polar cap shrinks rapidly with the onset of Martian spring and summer, and there are long periods when it cannot be seen at all except with large instruments. It may sometimes seem to vanish altogether. And, of course, the great dust-storms conceal surface details no matter what telescope is used.

For useful work, I consider that the minimum telescopic aperture is 8 inches for a reflector and 5 inches for a refractor. Other observers will disagree with me, and will say that a 6-inch reflector is adequate. They may well be right, but all I can do is to give a personal opinion. So far as magnification is concerned, I suggest that the proper procedure is to select the highest power which will give a really sharp image.

'Useful work' is a term which needs to be defined with regard to Mars. Now that we have the close-range space-probe pictures, there is little point in setting out to do ordinary mapping; the situation is much the same as for the Moon, where the pre-Space Age observers concentrated upon cartography, whereas the modern amateurs (and some professionals) are searching for what we may call time-dependent phenomena. There is, however, one very important difference. The Moon's structural features do not change, and have not done so for many millions of years. Mars has an atmosphere, and this atmosphere is dusty; the dark areas do show alterations in outline and intensity.

The Syrtis Major is a case in point. Sometimes it is said to look narrow, sometimes broader; and though many of these alleged changes can be put down to errors in observation or interpretation, it is at least possible that they have a basis of reality. Then there is Hellas, which can sometimes appear so brilliant that it is easily mistaken for an extra ice-cap, but at other oppositions is so obscure that it is hard to identify at all. Argyre shows variations in brightness of the same kind, though they are less striking.

Clouds are seen frequently, and observations are genuinely useful. Small clouds, occasionally well enough defined to have their positions determined, can shift from night to night, and can provide information about Martian wind velocities. The major dust-storms come into a different category, and here the best procedure is to do one's best to define their limits. They can spread with amazing rapidity, as happened in 1973. I observed the planet on 12 October, and the surface details were perfectly clear. The next three nights were cloudy. When I next observed, on 16 October, the dust had covered the planet, and I could see practically nothing at all.

Much depends upon the altitude of Mars. When the planet is low down a high magnification will be useless, and there is no alternative but to wait until the altitude has climbed to a respectable value. Also, it often happens that a very clear, transparent night will be unfavourable, with the disk of Mars wobbling like a jelly. Oddly enough, slight mist is sometimes advantageous, even though it cuts down the total light received, because the image can be pleasingly steady.

Without wanting to sound depressing, I must again stress that small telescopes are unable to show much on Mars even near opposition. There have been many published drawings made with such instruments, often showing fine details together with canals – all of which are completely spurious. This is no slur upon the integrity of the observers concerned, but the human eye is notoriously easy to deceive, and there is always the tendency to draw what one half-expects to see. Neither is this tendency confined to amateurs; far from it. One has only to look at the maps produced by Schiaparelli, Lowell and others, who were using powerful telescopes, but who still recorded the network of canals which we now know to be absolutely non-existent.

The phase should never be neglected when making a sketch. The fraction of the illuminated disk presented to us can go down to as little as 85 per cent, and if the observer merely draws a circle and fills in the detail he can see there are bound to be major errors. Drawing the disc to the correct phase can, however, be a tedious process. Personally, I admit that I 'cheat' by using prepared disks of the type shown on pages 24–25. They have the advantage of looking much neater than freehand phase drawings, and they are also shown against a black background, though I admit that this is not strictly necessary. To achieve absolute accuracy one ought to have a full set of disks, from 100 per cent (i.e. a perfect circle, as at opposition) down to 85 per cent (the minimum phase), but in practice I doubt whether one needs more than half a dozen – say 100 per cent, 98, 95, 92, 89, and 86. An error of one or two per cent is, frankly, unimportant.

In general it is wise to adopt a set scale for drawings; the Mars Section of the British Astronomical Association workers use 2 inches to the planet's full diameter, and this seems very suitable. Some observers vary the scale, drawing Mars largest when at opposition, but this seems an unnecessary refinement.

When sketching Mars, the first step is to survey the planet and see just what is on view. Then draw in the obvious details such as the polar caps and the main dark areas. When this has been done, check carefully and note the time (using the 24-hour clock, and ignoring Summer Time; always give G.M.T.). These main features should then be left unaltered. There is good reason for not changing them: Mars is rotating all the time, and the drift of the markings across the disk is perceptible even over periods of a few minutes.

Now change to the highest possible magnification and fill in the minor details, paying particular attention to the relative intensities of the various features, and concentrating upon anything which seems unusual (clouds in particular). Written notes, dealing with features of special interest, can be added, after which the whole drawing should be re-checked for accuracy. Finally, add the following data: date, time, name of observer, type and aperture of telescope, magnification, seeing conditions, and the longitude of the central meridian of Mars. Should any of these facts be omitted, the drawing promptly loses most of its value.

Seeing is usually given on the scale proposed by Antoniadi, ranging from I (perfect) through II (good), III (fair), IV (rather poor) down to V (so bad that one would not make a drawing at all unless there was some special reason for attempting it).

The longitude of the central meridian can be calculated easily, and involves nothing more frightening than simple addition or subtraction. Various publications, such as the *Handbook* of the British Astronomical Association, give the longitude of the central meridian for 0 hours G.M.T. each day, so that all one has to do is to allow for the interval between 0 hours and the time of observation. The longitude changes by the following amounts:

Hours	Change, degrees	Minutes	Change, degrees
1	14.6	1	0.2
2	29.2	2	0.5
3	43.9	3	0.7
4	58.5	4	1.0
5	73.1	5	1.2
6	87.7	6	1.5
7	102.3	7	1.7
8	117.0	8	2.0
9	131.6	9	2.2
10	146.2	10	2.4
		20	4.9
		30	7.3
		40	9.7
		50	12.2

Let me give a couple of examples from my own notebook

(1) I observed Mars on 2 September 1975, at 01.20 G.M.T., 15-in. reflector × 360; seeing III, phase 85%. Looking up the Handbook, I found that at 0 hours on 2 September the central longitude was 092.5.

Longitude at 0 hours:	92.5
+ 1 hour	14.6
+ 20 minutes	4.9
	——
Longitude at 01.20	112.0

This meant that the Syrtis Major was on the far side of the disk; but features such as Sirenum were on view – though since the diameter of the disk was only 9 seconds of arc, and the seeing was not good, I could make out little detail.

(2) Observation: 3 January 1976, at 16.27 G.M.T.; 15-in. refl. × 400, seeing II.

This time we have to subtract, because the observation was made 7 hours 33 minutes before 0 hours G.M.T.

Longitude of central meridian at 0 hours on 4 January: 036.7

We have to subtract for 7h 33m	7 hours:	102.3
	+30 minutes	7.3
	+ 3 minutes	0.7
		110.3

Since one cannot take 110.3 from 036.7, we must start by adding 360 degrees. 360 + 036.7 = 396.7.

Longitude at 0 hours on 4 January:	396.7
Subtract for 7h 33m:	110.3
	286.4

Therefore the Syrtis Major, whose longitude is approximately 290 degrees, was almost on the central meridian.

Because unconscious prejudice is the observer's greatest enemy, particularly when dealing with Mars, it is usually a good idea to leave working out the longitude of the central meridian until after the observation has been made – though I agree that this is rather pointless when some unmistakable feature is on view. In any case, avoid drawing any details which cannot be seen with absolute certainty.

Nowadays, many well-equipped amateurs make use of electronic equipment – mainly CCDs (Charge-Coupled Devices) – and produce images which rival the best images taken with large professional telescopes only a few years ago. Yet there is still a great amount of personal pleasure to be drawn from looking at the planet through the eye-end of a telescope, and recording what you can see.

APPENDIX II

Oppositions of Mars

The interval between successive oppositions of Mars is not constant; it may be as much as 810 days or as little as 764 days. Oppositions between 1900 and 2000 occurred on the following dates:

1901	Feb. 22	1926	Nov. 4	1952	Apr. 30	1978	Jan. 22
1903	Mar. 29	1928	Dec. 21	1954	June 24	1980	Feb. 25
1905	May 8	1931	Jan. 27	1956	Sept. 11	1982	Mar. 31
1907	July 6	1933	Mar. 1	1958	Nov. 17	1984	May 11
1909	Sept. 24	1935	Apr. 6	1960	Dec. 30	1986	July 10
1911	Nov. 25	1937	May 19	1963	Feb. 4	1988	Sept. 28
1914	Jan. 5	1939	July 23	1965	Mar. 9	1990	Nov. 27
1916	Feb. 9	1941	Oct. 10	1967	Apr. 15	1993	Jan. 7
1918	Mar. 15	1943	Dec. 5	1969	May 31	1995	Feb. 12
1920	Apr. 21	1946	Jan. 13	1971	Aug. 10	1997	Mar. 17
1922	June 10	1948	Feb. 17	1973	Oct. 25	1999	Apr. 24
1924	Aug. 23	1950	Mar. 23	1975	Dec. 15		

OPPOSITIONS OF 1999–2005

Date	Closest approach to Earth	Minimum distance, millions of miles	Max. apparent diameter, secs of arc	Max. magnitude	Constellation
1999 Apr. 24	1999 May 1	54	16.2	–1.5	Virgo
2001 June 13	2001 June 21	42	20.8	–2.1	Sagittarius
2003 Aug. 28	2003 Aug. 27	35	25.1	–2.7	Capricornus
2005 Nov. 7	2005 Oct. 30	43	20.2	–2.1	Aries

Mars is at perihelion on 25 Nov. 1999, 21 Oct. 2001, 30 Aug. 2003 and 17 July 2005; at aphelion on 2 Nov. 2000, 21 Sept. 2002 and 7 Aug. 2004.

Mars is at superior conjunction on 1 July 2000, 10 Aug. 2002 and 15 Sept. 2004. The greatest possible distance between Mars and Earth is 249 million miles. The minimum distance, as at the opposition of 1892, is 34.5 million miles.

Numerical Data

Distance from the Sun:

maximum	154,861,000 miles (1.666 a.u.)
mean	141,637,000 miles (1.524 a.u.)
minimum	128,412,000 miles (1.381 a.u.)

Axial rotation period (sol):   24 hours 37 minutes 22.7 seconds

Sidereal period: 686.980 days, or 668.60 sols

Orbital velocity, miles per second:

maximum	16.5
mean	15.0
minimum	13.6

Orbital inclination: $1°50'59''.4 (= 1°.9)$

Orbital eccentricity: 0.093

Mean synodic period: 779.74 days

Mean sidereal motion in 24 hours: 1886.52 seconds of arc

Axial inclination: 23°59'

Diameter: 4219 miles equatorial, 4196 miles polar

Apparent diameter: maximum 25''.7, minimum 3''.5

Maximum magnitude: −2.8

Albedo (mean):   0.16

Volume (Earth = 1):   0.1504

Mass (Earth = 1):   0.1074

Density (water = 1):   3.94

Surface area (Earth = 1):   0.28

Surface gravity (Earth = 1):   0.3799

Polar compression: 0.01

Escape velocity   3.1 miles per second

Mean diameter of Sun, as seen
from Mars: 21'

SATELLITE DATA

	Phobos	*Deimos*
Discoverer:	Hall, 16 Aug. 1877	Hall, 11 Aug. 1877
Mean distance from centre of Mars:	5825 miles	14,575 miles
Sidereal period:	7h 39m 13s.85	30h 17m 54s.87
	(0.319 days)	(1.262 days)
Mean synodic period:	7h 39m 26s.6	30h 21m 15s.7
Orbital inclination:	1.1 degrees	1.8 degrees
Orbital eccentricity:	0.0210	0.0028
Mean angular distance from Mars, at mean opposition	24.6 seconds of arc	61.8 seconds of arc
Mean visual opposition magnitude:	11.6	12.8
Diameter in miles:	16.8 × 13.7 × 11.2	9.3 × 7.5 × 6.2
Escape velocity, miles per second:	0.0099	0.0049
Mean apparent magnitude, seen from Mars:	–3.9	–0.1
Density (water = 1):	2.0	1.7

APPENDIX IV

Space Missions to Mars

Name	Country	Launch date	Encounter date	Closest approach of orbiter, miles	Landing site	Comments
Mars 1	U.S.S.R.	1 Nov. 1962	?6 June 1963	?120,000	–	Contact lost on 21 Mar. 1963, at 66,000 miles.
Mariner 3	U.S.A.	5 Nov. 1964	–	–	–	Shroud failure. In solar orbit, but contact lost soon after launch.
Mariner 4	U.S.A.	28 Nov. 1964	15 July 1965	6084	–	Returned 21 images, plus data. Contact lost on 20 Dec. 1967. Now in solar orbit.
Zond 2	U.S.S.R.	30 Nov. 1964	?6 Aug. 1965	?930	–	Contact lost on 2 May 1965.
Zond 3	U.S.S.R.	18 July 1965	?	?	–	Contact lost soon after launch.
Mariner 6	U.S.A.	24 Feb. 1969	31 July 1969	2130	–	Returned 76 images. Flew over Martian equator. In solar orbit.
Mariner 7	U.S.A.	27 Mar. 1969	4 Aug. 1969	2130	–	Returned 126 images, mainly over S. hemisphere. In solar orbit.
Mariner 8	U.S.A.	8 May 1971	–	–	–	Total failure; fell in the sea.
Mars 2	U.S.S.R.	19 May 1971	27 Nov. 1971	In orbit, 1520 × 15,165	44 S, 213 W	Landed in Eridania. Carried Soviet pennant, but no images received.
Mars 3	U.S.S.R.	28 May 1971	2 Dec. 1971	In orbit, 965 × 132,250	45 S, 158 W	Landed in Phaethontis. Contact with lander lost 20 seconds after arrival.

Name	Country					Notes
Mariner 9	U.S.A.	30 May 1971	13 Nov. 1971	In orbit, 1019 × 10,440	–	Returned 7329 images. Contact lost on 27 Oct. 1972.
Mars 4	U.S.S.R.	21 July 1973	10 Feb. 1974	Over 1300	–	Missed Mars; some data returned during fly-by. Now in solar orbit.
Mars 5	U.S.S.R.	25 July 1973	10 Feb. 1974	In orbit, 1093 × 20,200	–	Failure; contact lost.
Mars 6	U.S.S.R.	5 Aug. 1973	12 Mar. 1974	–	?24 S, 25 W	Landed in Erythræum, 3 Dec. 1973, but contact lost during landing sequence.
Mars 7	U.S.S.R.	9 Aug. 1973	9 Mar. 1974	795	–	Missed Mars; failed to orbit.
Viking 1	U.S.A.	20 Aug. 1975	19 June 1976	In orbit	23.4 N, 47.5 W	Landed in Chryse, 20 July 1976. Contact lost on 13 Nov.
Viking 2	U.S.A.	9 Sept. 1975	7 Aug. 1976	In orbit	48 N, 226 W	Landed in Utopia. 3 Sept. 1976. Contact lost on 4 Dec.
Phobos-1	U.S.S.R.	7 July 1988	?	?	–	Contact lost on 31 Aug. 1988.
Phobos-2	U.S.S.R.	12 July 1988	29 Jan. 1989	In orbit, 516 × 49,560	–	Contact lost on 27 Mar. 1989. Some images and data from Mars and Phobos returned.
Mars Observer	U.S.A.	25 Sept. 1992	24 Aug. 1993	?	–	Contact lost on 21 Aug. 1993. Orbit unknown.
Mars 96	Russia	16 Nov. 1996	–	–	–	Total failure; fell in the sea.
Pathfinder	U.S.A.	4 Dec. 1996	4 July 1997	–	19.33 N, 33.55 W	Landed in Ares Vallis, 4 July 1997. Carried Sojourner rover.
Global Surveyor	U.S.A.	7 Nov. 1996	11 Sept. 1997	In orbit	–	
Nozomi	Japan	3 July 1998	–	–	–	Intended orbiter.

FUTURE MISSIONS TO MARS

Many new missions to Mars will be launched in the near future. The following list was compiled in July 1998; no doubt it will be drastically modified.

Name	Country	Estimated launch	Estimated arrival	Purpose
Mars Surveyor 98 (orbiter)	U.S.A.	Dec. 1998	1999	Orbiter; mapper, relay.
Mars Surveyor 98 (lander)	U.S.A.	4 Jan. 1999	1999	Lander in S. hemisphere. Use Mars 98 as a relay.
Mars Surveyor 2001 (orbiter)	U.S.A.	Mar. 2001	2001	Orbiter; gamma-ray spectrometer.
Mars Surveyor 2001 (lander)	U.S.A.	Apr. 2001	2001	Lander; carries a rover. Use Surveyor 2001 as a relay.
Mars 2001	Russia	2001	2001	Lander; carries a rover.
Mars Surveyor 2003	U.S.A.	2003	2003	Orbiter to relay data from lander/rover.
Mars Express 2003	Europe	2003	2003	Orbiter.
Mars Surveyor 2005	U.S.A.	2005	2005	Sample-and-return mission.

Martian Albedo Features

I give here a list of 'albedo features', with the old names, to go with the south-at-the-top chart, because these are the details which will normally be seen by the observer with a telescope of moderate aperture.

Name	Central lat.	Long. W	Name	Central lat.	Long. W
Mare Acidalium	45.0 N	30.0	Coprates	15.0 S	65.0
Aeolis	5.0 S	215.0	Cyclopia	5.0 S	230.0
Aeria	10.0 N	310.0	Cydonia	40.0 N	0.0
Aetheria	40.0 N	230.0	Deltoton Sinus	4.0 S	305.0
Aethiopis	10.0 N	230.0	Deucalionis Regio	15.0 S	340.0
Amazonis	0.0	140.0	Deuteronilus	35.0 N	0.0
Amenthes	5.0 N	250.0	Diacria	50.0 N	180.0
Aonium Sinus	45.0 S	105.0	Edom	0.0	345.0
Arabia	20.0 N	330.0	Electris	45.0 S	190.0
Arcadia	45.0 N	100.0	Elysium	25.0 N	210.0
Argyre	45.0 S	25.0	Eridania	45.0 S	220.0
Arnon	48.0 N	335.0	Mare Erythræum	25.0 S	40.0
Auroræ Sinus	15.0 S	50.0	Eunostos	22.0 N	120.0
Ausonia	40.0 S	250.0	Euphrates	20.0 N	335.0
Mare Australe	60.0 S	20.0	Gehon	15.0 N	0.0
Baltia	60.0 N	50.0	Mare Hadriacum	40.0 S	270.0
Boerosyrtis	55.0 N	290.0	Hellas	40.0 S	290.0
Mare Boreum	60.0 N	180.0	Depressio		
Candor	3.0 N	75.0	Hellespontica	60.0 S	340.0
Casius	40.0 N	260.0	Hellespontus	50.0 S	325.0
Cebrenia	50.0 N	210.0	Hesperia	20.0 S	240.0
Cecropia	60.0 N	320.0	Hiddekel	15.0 N	345.0
Ceraunius	20.0 N	93.0	Hyperboreus Lacus	75.0 N	60.0
Cerberus	15.0 N	205.0	Iapygia	20.0 S	295.0
Chalce	50.0 S	0.0	Icaria	40.0 S	130.0
Chersonesus	50.0 S	260.0	Isidis Regio	20.0 N	275.0
Mare Chronium	58.0 S	210.0	Ismenius Lacus	40.0 N	330.0
Chryse	10.0 N	30.0	Jamuna	10.0 N	40.0
Chrysokeras	50.0 S	110.0	Juventæ Fons	5.0 S	63.0
Mare Cimmerium	20.0 S	220.0	Læstrygon	0.0	200.0
Claritas	35.0 S	110.0	Lemuria	70.0 N	200.0
Copaïs Palus	55.0 N	280.0	Libya	0.0	270.0

Name	Central lat.	Long. W	Name	Central lat.	Long. W
Lunæ Palus	15.0 N	65.0	Protonilus	42.0 N	315.0
Margaritifer Sinus	10.0 S	25.0	Pyrrhæ Regio	15.0 S	38.0
Memnonia	20.0 S	150.0	Sinus Sabæus	8.0 S	340.0
Sinus Meridiani	5.0 S	0.0	Scandia	60.5 N	147.9
Meroe	35.0 N	285.0	Mare Serpentis	30.0 S	320.0
Moab	20.0 N	350.0	Sinai	20.0 S	70.0
Mœris Lacus	8.0 N	270.0	Mare Sirenum	30.0 S	155.0
Nectar	28.0 S	72.0	Sithonius Lacus	45.0 N	245.0
Neith Regio	38.0 N	272.0	Solis Lacus	28.0 S	90.0
Nepenthes	20.0 N	260.0	Styx	30.0 N	200.0
Nereidum Fretum	45.0 S	55.0	Syria	20.0 S	100.0
Niliacus Lacus	30.0 N	30.0	Syrtis Major	10.0 N	290.0
Nilokeras	30.0 N	55.0	Tanaïs	50.0 N	70.0
Nilosyrtis	42.0 N	290.0	Tempe	40.0 N	70.0
Nix Olympica	20.0 N	130.0	Tharsis	0.0	0.0
Noachis	45.0 S	330.0	Thaumasia	35.0 S	85.0
Ogygis Regio	45.0 S	65.0	Thoth	30.0 N	255.0
Olympia	80.0 N	200.0	Thyle I, II	70.0 S	180.0
Ophir	10.0 S	65.0	Thymiamata	10.0 N	10.0
Ortygia	60.0 N	0.0	Tithonius Lacus	5.0 S	85.0
Oxia Palus	20.0 N	180.0	Tractus Albus	30.0 N	80.0
Panchaïa	60.0 N	200.0	Trinacria	25.0 S	268.0
Pandoræ Fretum	25.0 S	316.0	Trivium Charontis	20.0 N	198.0
Phæthontis	50.0 S	155.0	Mare Tyrrhenum	20.0 S	255.0
Phison	20.0 N	320.0	Uchronia	70.0 N	260.0
Phlegra	31.3 N	187.8	Umbra	50.0 N	290.0
Phœnicis Lacus	12.0 S	110.0	Utopia	50.0 N	250.0
Phrixi Regio	40.0 S	70.0	Vulcani Pelagus	35.0 S	15.0
Promethei Sinus	65.0 S	280.0	Xanthe	10.0 N	50.0
Propontis	45.0 N	185.0	Yaonis Regio	40.0 S	320.0
Protei Regio	23.0 S	50.0	Zephyria	0.0	195.0

NAMES OF ALBEDO FEATURES

The classical albedo nomenclature closely followed Schiaparelli's. The space research results have led to major modifications, but it may be of interest to give some of the old derivations.

MARE ACIDALIUM	Acidalia fountain in Bœotia, where the Graces bathed.
AEOLIS	Floating island where the winds were caged.
AERIA	Greek name for Egypt.
AETHERIA	Upper world: land of the living.
AETHIOPIS	Country of the Ethiopians.
AMAZONIS	Land of the Amazons.
AONIUM SINUS	The Aonides, or Muses.
ARABIA	Terrestrial Arabia.

ARCADIA	Mountainous region in Southern Greece.
ARGYRE	'Silver' island at the mouth of the River Ganges.
AURORÆ SINUS	Bay of Dawn.
AUSONIA	Country of the Aruncii (Ausones in Greek).
MARE AUSTRALE	Southern Sea.
BALTIA	North European island.
MARE BOREUM	Northern Sea.
CANDOR	'White' in Latin.
CASIUS	Two sanctuaries of Zeus.
CEBRENIA	Main country of the Trojan plain.
CERAUNIUS	'Thunderclap'. Ceraunii Mountains in Greece.
CERBERUS	Three-headed dog who guarded the gates to the Underworld.
CHERSONESUS	Gallipoli Peninsula.
MARE CHRONIUM	Cronian Sea, where dead calm prevails.
CHRYSE	The Golden Plain.
MARE CIMMERIUM	Named for the Cimmerians, ancient Thracian seafarers.
CLARITAS	Latin, 'Bright'.
COPAÏS PALUS	Marsh north of Mount Helicon, in Greece.
COPRATES	Old name for Persian river.
CYCLOPIA	Land of the Cyclops.
CYDONIA	Crete.
DELTATON SINUS	'Triangle Bay'. (Triangle with Iapygia and Œnotria.)
DEUCALIONIS REGIO	Named for Deucalion, King of Thessaly.
DIACRIA	Upland area in Northern Greece.
DIOSCURIA	Old name for Sparta.
EDOM	Biblical name of the country of the Edomites, south of Judah.
ELECTRIS	Island famous for amber.
ELYSIUM	Home of the blessed, on the western edge of the world.
ERIDANIA	Region of the River Po, in Italy.
MARE ERYTHRÆUM	Indian Ocean.
EUNOSTOS	Infernal regions beyond Elysium.
EUPHRATES, GEHON	Rivers.
MARE HADRIACUM	Adriatic Sea.
HELLAS	Greece.
HESPERIA	'The Occident' (Italy or Spain).
HIDDEKEL	River Tigris.
IAPYGIA	The Salantine Peninsula.
ICARIA	Crete.
ISIDIS REGIO	Named for the Egyptian goddess Isis.
ISMENIUS LACUS	Thebes.
JUVENTÆ FONS	Fountain of Youth.
LÆSTRYGON	Giants who lived in the West.
LEMURIA	Old submerged continent.
LIBYA	Terrestrial Libya.
LUNÆ PALUS	Marsh of the Moon.
MARGARITIFER SINUS	Bay of Pearls.
MEROE	Atbar Island, on the Nile.
MŒRIS LACUS	Lake in the Libyan Desert.

NECTAR	Drink of the Olympians.
NERPENTHES	Egyptian drug of forgetfulness.
NEREIDUM FRETUM	Strait of the Nereids (sea-nymphs).
NILIACUS LACUS	Lake of the Nile.
NILOKERAS	'Horn of the Nile'.
NIX OLYMPICA	Olympic Snow.
NOACHIS	Noah's country.
OGYGIS REGIO	Named for Ogyges, King of Thebes or Athens.
OPHIR	Land to which Solomon sent an expedition; India?
ORTYGIA	Delos.
OXIA PALUS	Sea of Aral.
PANCHAÏA	Island in the Red Sea.
PANDORÆ FRETUM	Named for Pandora, who opened a box and let loose all the evils of the world.
PHÆTHONTIS	Named for Phæthon, the boy who recklessly drove the Sun-chariot.
PHLEGRA	'Burning Plain' in the Chalcidian Peninsula, Greece.
PHŒNICIS LACUS	Lake of the Phœnix.
PROPONTIS	The Sea of Marmora.
PROTONILUS	Part of the Nile.
PYRRHÆ REGIO	Named for Pyrrha, wife of Deucalion.
SINUS SABÆUS	The Red Sea.
SCANDIA	Southern Scandinavia.
MARE SERPENTIS	The Serpent Sea.
MARE SIRENUM	Sea of the Sirens.
SITHONIUS LACUS	Thrace.
SOLIS LACUS	Lake of the Sun.
STYX	River bordering the Underworld.
SYRIA	Terrestrial Syria.
SYRTIS MAJOR	Libyan Gulf; now Gulf of Sirte.
TANAÏS	River Don (Russia).
TEMPE	Valley south of Mount Olympus, in Greece.
THARSIS	Tartessus; Spanish town connecting East and West.
THAUMASIA	Named for Thaumas, the god of clouds.
THOTH	Egyptian messenger god.
THYLE	Thule; possibly Norway.
TITHONIUS LACUS	Named for Tithonius, who was granted eternal life but not eternal youth.
TRACTUS ALBUS	'White tract' (Latin).
TRINACRIA	Sicily.
MARE TYRRHENUM	Tyrrhenian Sea.
UMBRA	Land of shadow.
UTOPIA	Ideal state.
VULCANI PELAGUS	Named for Vulcan, the blacksmith of the gods.
XANTHE	Golden yellow land.
YAONIS REGIO	Named for the Chinese emperor Yao.
ZEPHYRIA	Land of the West Wind.

Features on Mars

The features shown here are not observable with ordinary Earth-based telescopes, with a few exceptions; the main dark areas, and some of the smaller features such as Olympus Mons, formerly Nix Olympica (the Olympic Snow). Positions refer to the centres of features.

CATENA (chain of craters)

	Lat.	Long. W	Length, miles
Coprates Catena	15.1 S	59.4	151
Elysium Catena	18.0 N	210.4	137
Tithoniæ Catena	5.5 S	71.5	236
Tractus Catena	27.9 N	103.2	767

CHASMA (large rectilinear chain)

	Lat.	Long. W	Length, miles
Chasma Australe	82.9 S	273.8	305
Chasma Boreale	83.2 N	21.3	198
Capri Chasma	8.7 S	42.6	931
Candor Chasma	6.5 S	71.0	507
Eos Chasm	12.6 S	45.1	599
Gangis Chasma	8.4 S	48.1	536
Hebes Chasma	1.1 S	76.1	177
Ius Chasma	7.2 S	84.6	623
Juventæ Chasma	1.9 S	61.8	308
Melas Chasma	10.5 S	72.9	327
Ophir Chasma	4.0 S	72.5	156
Tithonium Chasma	4.6 S	86.5	562

CRATERS

	Lat.	Long. W	Diameter, miles
Adams	31.3 N	197.1	62
Antoniadi	21.7 N	299.0	237
Arago	10.5 N	330.2	96
Baldet	23.0 N	94.5	121
Barabashov	47.6 N	68.5	78

Barnard	61.3 S	298.4	80
Becquerel	22.3 N	7.9	104
Beer	14.6 S	8.2	50
Bjerknes	43.4 S	188.7	55
Boeddicker	14.8 S	197.6	67
Bond	33.3 S	35.7	65
Brashear	54.1 S	119.2	78
Burton	14.5 S	156.3	85
Byrd	65.6 S	231.9	76
Campbell	54.0 S	195.0	76
Cassini	23.8 N	327.9	258
Cerulli	32.7 N	337.9	74
Clark	55.7 S	133.2	58
Coblentz	55.3 S	90.2	69
Comas Solà	20.1 S	158.4	82
Copernicus	50.0 S	168.6	181
Curie	29.2 N	4.9	61
Darwin	57.2 S	19.2	103
Douglass	51.7 S	70.4	60
Eddie	12.5 N	217.8	56
Eiriksson	19.6 S	173.7	35
Escalante	0.3 N	244.8	52
Fesenkov	21.9 N	86.4	53
Flammarion	25.7 N	311.7	99
Flaugergues	17.0 S	340.9	146
Focas	33.9 N	347.2	51
Fournier	4.2 S	287.5	70
Galle	50.8 S	30.7	143
Gill	15.8 N	354.5	50
Gledhill	53.5 S	272.9	45
Graff	21.5 S	206.0	98
Green	42.4 S	8.3	114
Hale	36.1 S	36.3	84
Halley	48.6 S	59.2	50
Hartwig	38.7 S	15.6	65
Helmholtz	45.6 S	21.1	67
Henry	11.0 N	336.8	103
Herschel	14.9 S	230.1	189
Hipparchus	45.0 S	151.1	65
Holden	26.5 S	33.0	88
Holmes	75.0 S	293.9	68
Hooke	45.0 S	44.4	90
Huggins	49.3 S	204.3	51
Hussey	53.8 S	126.5	62
Huygens	14.0 S	304.4	283
Janssen	2.8 N	322.4	96
Jones	19.1 S	19.8	53

Kaiser	46.6 S	340.9	125
Keeler	60.7 S	151.2	57
Kepler	47.2 S	218.7	136
Knobel	6.6 S	226.9	79
Korolev	72.9 N	195.8	84
Kunowsky	57.0 N	9.0	37
Lampland	36.0 S	79.5	44
Lassell	21.0 S	62.4	53
Le Verrier	38.2 S	342.9	86
Liu Hsin	53.7 S	171.4	80
Lockyer	28.2 N	199.4	46
Lohse	43.7 S	16.4	97
Lomonosov	64.8 N	8.8	94
Lowell	52.3 S	81.3	125
Lyot	50.7 N	330.7	137
Main	76.8 S	310.9	63
Maraldi	62.2 S	32.1	74
Mariner	35.2 S	164.3	94
Martz	35.2 S	215.8	57
McLaughlin	22.1 N	22.5	56
Mie	48.6 N	220.4	58
Milankovič	54.8 N	146.6	70
Moreux	42.2 N	315.5	86
Nansen	50.5 S	140.3	51
Newton	40.8 S	157.9	178
Nicholson	0.1 N	164.5	71
Niesten	28.2 S	302.1	71
Oudemans	10.0 S	91.7	75
Pasteur	19.6 N	335.5	71
Perepelkin	52.8 N	64.6	70
Pickering	34.4 S	132.8	70
Playfair	78.0 S	125.5	42
Porter	50.8 S	113.8	70
Proctor	47.9 S	330.4	104
Ptolemæus	46.4 S	157.5	114
Quenisset	34.7 N	319.4	79
Renaudot	42.5 N	297.4	42
Ritchey	28.9 S	50.9	51
Ross	57.6 S	107.6	55
Russell	55.0 S	347.4	86
Schiaparelli	2.5 S	343.4	287
Schröter	1.8 S	303.6	209
Secchi	57.9 S	257.7	135
Sharonov	27.3 N	58.5	59
Slipher	47.7 S	84.5	80
South	77.0 S	338.0	69
Stokes	56.0 N	189.0	44

Tikhov	51.2 S	254.1	67
Trouvelot	16.3 N	13.0	104
Trümpler	61.7 S	50.6	48
Tycho Brahe	49.5 S	213.8	67
Vishniac	76.7 S	276.1	47
Vogel	15.1 S	5.9	26
von Kármán	64.3 S	58.4	62
Wallace	52.8 S	249.2	99
Wegener	64.3 S	4.0	44
Wirtz	48.7 S	25.8	80
Wislencius	18.3 S	348.7	66
Zulanka	2.3 S	42.3	29

FOSSA (ditch)

	Lat.	Long. W	Length, miles
Alba Fossæ	43.4 N	103.6	1290
Ceraunius Fossæ	24.8 N	110.5	442
Claritas Fossæ	34.8 S	99.1	1264
Elysium Fossæ	27.5 N	219.5	692
Hephæstus Fossæ	20.5 N	237.5	328
Mareotis Fossæ	45.0 N	79.3	494
Memnonia Fossæ	21.9 S	154.4	651
Nili Fossæ	24.0 N	283.0	441
Sirenum Fossæ	34.5 S	158.2	1686
Tantalus Fossæ	44.5 N	102.4	1237
Tempe Fossæ	40.2 N	74.5	965
Thaumasia Fossæ	45.9 S	97.3	695
Tithoniæ Fossæ	1.7 S	81.9	341
Zephyrus Fossæ	23.9 N	214.4	281

LABYRINTHUS (valley complex)

	Lat.	Long. W	Diameter, miles
Noctis Labyrinthus	7.2 S	101.3	610

MENSA (mesa)

	Lat.	Long. W	Diameter, miles
Aeolis Mensæ	3.7 S	218.5	523
Cydonia Mensæ	37.0 N	12.8	531
Deuteronilus Mensæ	45.7 N	337.9	390
Nepenthes Mensæ	9.3 N	241.7	1059
Nilosyrtis Mensæ	35.4 N	293.6	431
Protonilus Mensæ	44.2 N	309.4	368
Zephyria Mensæ	10.1 S	188.1	143

MONS (mountain or volcano)

	Lat.	Long. W	Base diameter, miles
Arsia Mons	9.4 S	120.5	301
Ascræus Mons	11.3 N	104.5	287
Charitum Montes	58.3 S	44.2	878
Elysium Mons	25.0 N	213.0	268
Hellespontus Montes	45.5 S	317.0	423
Libya Montes	2.7 N	271.2	763
Nereidum Montes	41.0 S	43.5	1042
Olympus Mons	18.4 N	133.1	388
Pavonis Mons	0.3 N	112.8	233
Phlegra Montes	40.9 N	197.4	814
Tartarus Montes	25.1 N	188.7	628
Tharsis Montes	2.8 N	113.3	1308

PATERA (saucer-shaped, volcanic structure)

	Lat.	Long. W	Diameter, miles
Alba Patera	40.5 N	109.9	288
Amphitrites Patera	59.1 S	299.0	86
Apollinaris Patera	8.3 S	186.0	123
Biblis Patera	2.3 N	123.8	73
Diacria Patera	34.8 N	132.7	46
Hadriaca Patera	30.6 S	267.2	280
Orcus Patera	14.4 N	181.5	237
Tyrrhena Patera	21.9 S	253.2	371
Ulysses Patera	2.9 N	121.5	70
Uranius Patera	26.7 N	92.6	172

PLANITIA (smooth plain)

	Lat.	Long. W	Diameter, miles
Acidalia Planitia	54.6 N	19.9	1735
Amazonis Planitia	16.0 N	158.4	1750
Arcadia Planitia	46.4 N	152.1	1897
Argyre Planitia	46.4 S	42.8	539
Chryse Planitia	27.0 N	36.0	932
Elysium Planitia	14.3 N	241.1	2423
Hellas Planitia	44.3 S	293.8	1564
Isidis Planitia	13.1 N	271.0	769
Utopia Planitia	47.6 N	277.3	2036

PLANUM (smooth plain)

	Lat.	Long. W	Diameter, miles
Auroræ Planum	11.1 S	50.2	367
Australe Planum	80.3 S	155.1	816

Boreum Planum	85.0 N	180.0	663
Bosporos Planum	33.4 S	64.0	205
Chronium Planum	62.0 S	212.6	871
Hesperia Planum	18.3 S	251.6	1161
Icaria Planum	42.7 S	107.2	522
Lunæ Planum	9.6 N	66.6	1147
Malea Planum	65.9 S	297.4	664
Ophir Planum	9.6 S	62.2	664
Sinai Planum	12.5 S	87.1	661
Solis Planum	20.9 S	94.6	910
Syria Planum	12.0 S	103.9	470
Syrtis Major Planum	9.5 N	289.6	843
Thaumasia Planum	22.0 S	65.0	578

TERRA (land)

	Lat.	*Long. W*	*Diameter, miles*
Aonia Terra	58.2 S	94.8	2095
Arabia Terra	25.0 N	330.0	3729
Cimmeria Terra	34.0 S	215.0	1420
Margaritifer Terra	16.2 S	21.3	3365
Meridiani Terra	7.2 S	356.0	1008
Noachis Terra	45.0 S	350.0	2175
Promethei Terra	52.9 S	262.2	1716
Sabæa Terra	10.4 S	330.6	850
Sirenum Terra	37.0 S	160.0	1345
Tempe Terra	41.3 S	70.5	1277
Tyrrhena Terra	14.3 S	278.5	1751
Xanthe Terra	4.2 N	46.3	1911

THOLUS (domed hill)

	Lat.	*Long. W*	*Base diameter, miles*
Albor Tholus	19.3 N	209.8	103
Ceraunius Tholus	24.2 N	97.2	67
Hecates Tholus	32.7 N	209.8	113
Iaxartes Tholus	72.0 N	15.0	33
Jovis Tholus	18.4 N	117.5	38
Tharsis Tholus	13.4 N	90.8	95
Uranius Tholus	26.5 N	97.8	46

VALLIS (valley)

	Lat.	*Long. W*	*Length, miles*
Al-Qahira Vallis	17.5 S	196.7	339
Ares Vallis	9.7 N	23.4	1050
Auqakuh Vallis	28.6 N	299.3	121

Huo Hsing Vallis	31.5 N	293.9	211
Kasei Vallis	22.8 N	68.2	1380
Ma'adim Vallis	21.0 S	182.7	535
Maja Vallis	15.4 N	56.7	814
Mangala Vallis	7.4 S	150.3	547
Marti Vallis	11.0 N	182.0	1057
Nirgal Vallis	28.3 S	41.9	318
Shalbatana Vallis	5.6 N	43.1	417
Simud Vallis	11.5 N	38.5	667
Tiu Vallis	8.6 N	34.8	603
Valles Marineris	11.6 S	70.0	2566

VASTITAS (extensive plain)

	Lat.	*Long. W*	*Diameter, miles*
Vastitas Borealis	67.5 N	180.0	6214

Further Reading

BAKER, V. *The Channels of Mars*, Hiller, 1982
BLUNCK, J. *Mars and its Satellites*, Exposition Press, 1979 (Nomenclature)
CARR, M. R. *The Surface of Mars*, Yale University Press, 1982
CATTERMOLE, P. *Mars*, Chapman & Hall, 1992
HOYT, W. *Lowell and Mars*, University of Arizona Press, 1976
MOORE, P. *Atlas of the Universe*, George Philip, 1998
———. *Mission to the Planets*, Cassell, 1997

Index

Surface features are indicated by an asterisk.

217